Reviews and critical articles covering the entire field of normal anatomy (cytology, histology, cyto- and histochemistry, electron microscopy, macroscopy, experimental morphology and embryology and comparative anatomy) are published in Advances in Anatomy, Embryology and Cell Biology. Papers dealing with anthropology and clinical morphology that aim to encourage co-operation between anatomy and related disciplines will also be accepted. Papers are normally commissioned. Original papers and communications may be submitted and will be considered for publication provided they meet the requirements of a review article and thus fit into the scope of "Advances". English language is preferred, but in exceptional cases French or German papers will be accepted.

It is a fundamental condition that submitted manuscripts have not been and will not simultaneously be submitted or published elsewhere. With the acceptance of a manuscript for publication, the publisher acquires full and exclusive copyright for all languages and countries.

Twenty-five copies of each paper are supplied free of charge.

Manuscripts should be addressed to

Prof. Dr. F. **BECK,** Department of Anatomy, University of Leicester, 6 University Road, GB-Leicester LE1 7RH

Prof. W. **HILD,** Department of Anatomy, Medical Branch, The University of Texas, Galveston, Texas 77550/USA

Prof. Dr. R. **ORTMANN,** Anatomisches Institut der Universität, Lindenburg, D-5000 Köln-Lindenthal

Prof. J.E. **PAULY**, Department of Anatomy, University of Arkansas for Medical Sciences, Little Rock, Arkansas 72205/USA

Prof. Dr. T.H. **SCHIEBLER,** Anatomisches Institut der Universität, Koellikerstraße 6, D-8700 Würzburg

Advances in Anatomy
Embryology and Cell Biology

Vol. 92

Editors
F. Beck, Leicester W. Hild, Galveston
R. Ortmann, Köln J.E. Pauly, Little Rock
T.H. Schiebler, Würzburg

Jan J. Heimans   Jaap Valk   Anthony H.M. Lohman

Angiographic Anatomy of the Anterior Inferior Cerebellar Artery

With 69 figures

Springer-Verlag
Berlin Heidelberg NewYork Tokyo 1985

Jan J. Heimans, M.D., Ph.D.
Jaap Valk, M.D., Ph.D.
Anthony H.M. Lohmann, M.D., Ph.D.

Free University Hospital
Amstelveenseweg 573, P.O. Box 7057
NL-1007 MB Amsterdam

Library of Congress Cataloging in Publication Data
Heimans, Jan J. Angiographic anatomy of the anterior inferior cerebellar artery.
(Advances in anatomy, embryology and cell biology; vol. 92)
Bibliography; p. Includes index. 1. Cerebellum – Tumors – Diagnosis. 2. Cerebellum – Blood-vessels –
Anatomy. 3. Cerebellum – Blood-vessels – Radiography. 4. Arteries – Anatomy. I. Valk, J. II. Lohman,
A.H.M. III. Title. IV. Series: Advances in anatomy, embryology, and cell biology; v. 92.
[DNLM: 1. Arteries – anatomy & histology. 2. Cerebellum – blood supply. 3. Cerebral Angiography.
W1 AD433K v. 92 / WL 320 H467a]
QL801.E67 vol. 92   574.4 s   84-23548   [RC280.B7]   [616.99′281]
ISBN-13: 978-3-540-13768-9   e-ISBN-13: 978-3-642-70023-1
DOI: 10.1007/ 978-3-642-70023-1

Typesetting,: Universitätsdruckerei H. Stürtz AG, Würzburg
2121/3140-543210

Contents

Abbreviations

AICA	anterior inferior cerebellar artery
AP	anteroposterior
BA	basilar artery
CMB	caudomedial branch
CPA	cerebellopontine angle
CT	computed tomography
HB	hemispheric branch
HF	horizontal fissure
IAA	internal auditory artery
IAccA	inferior accessory AICA
MB	marginal branch
ML	meatal loop
PICA	posterior inferior cerebellar artery
PLF	posterolateral fissure
RLB	rostrolateral branch
SAccA	superior accessory AICA
SCA	superior cerebellar artery
VA	vertebral artery

1 INTRODUCTION

The anterior inferior cerebellar artery (AICA) is one of the major branches of the basilar artery and supplies part of the pons, the upper medulla, and the cerebellar hemisphere. The artery can be visualized by means of vertebral angiography. This technique of examination was carried out for the first time in 1933 by Moniz and co-workers (Moniz and Alves 1933, Moniz et al. 1933). During the decades that followed, angiographic techniques improved considerably, with the result that more details of the angioarchitecture of the posterior cranial fossa could be demonstrated. Satisfactory visualization of the AICA and its branches depends greatly on the use of subtraction, and this is the reason why detailed reports on the angiographic appearance of the artery were for the greater part published after 1965, when subtraction techniques were more consistently used (Takahashi et al. 1968, 1974; Gerald et al. 1973).

The angiographic appearance of the various segments of the AICA in the lateral projection, both in the normal situation and in the presence of tumors, has been studied by Naidich et al. (1976a, b). The primary aim was to recognize and denominate the separate branches, loops, and segments of the AICA in order to locate tumors on the basis of displacements of portions of the artery. The fact that the course, caliber, and distribution of the AICA are very variable was not emphasized.

Since the introduction of computed tomography (CT), the localizing role of vertebral angiography is of less importance. However, in cases in which operation on posterior fossa tumors is considered, angiography offers essential information by revealing both the vascularization of the tumor and the relation between the tumor and major arteries, thus reducing the chance of damage to the AICA which may occur during cerebellopontine angle surgery. Such damage may result in lateral pontine infarction.

In the last few years, several reports have been published on extra-intracranial bypass operations in patients suffering from occlusive disease in the vertebrobasilar system (Khodadad et al. 1977; Sundt and Piepgras 1978; Ausman et al. 1979, 1981a, b). In these cases, also, thorough knowledge of the variations in the course and distribution of the posterior fossa arteries is of great practical importance.

The aim of the present study is to describe the variations of the AICA and to identify its types of appearance in normal angiograms as well as in angiograms of patients suffering from posterior fossa tumors or from ischemic lesions in the vertebrobasilar territory. For this purpose, an anatomic study of 20 normal specimens was undertaken. Four main types of the AICA could be distinguished.

For the radiological part of the study, one hundred normal vertebral angio-

1

grams, made between 1976 and 1982 in the Valeriuskliniek and the Academisch Ziekenhuis der Vrije Universiteit (Amsterdam, the Netherlands), were reviewed. The AICAs were classified in the same way as in the anatomic study. This classification was also used in the analysis of 41 vertebral angiograms of patients with posterior fossa tumors and nine angiograms of patients with ischemic disturbances in the posterior cranial fossa.

2 ANATOMY

2.1 Review of the Literature

2.1.1 General Description of the Arteries of the Posterior Cranial Fossa

Two vertebral arteries (VAs) enter the skull through the foramen magnum. Ventrobasal to the brain stem, both arteries pass to the lower border of the pons, where they unite to form the basilar artery (BA). Exceptionally, the VAs do not unite, and the BA is then formed by one of them (Berry and Andersen 1910; McMinn 1953; Walker 1965). The caliber of the VAs varies widely, and in most cases the right and left arteries are of different sizes. Further, the junction of the VAs is not always exactly at the pontomedullary sulcus. The vessels may unite caudal to this sulcus, in which case the BA is longer than usual. In contrast, a short BA occurs when the VAs unite rostral to the ponto-medullary sulcus. The BA runs in the cisterna pontis and extends rostrally as far as the upper border of the pons, where in most cases it bifurcates, forming the posterior cerebral arteries. Rarely, only one posterior cerebral artery is given off. The two VAs and the BA give rise to the arteries supplying the brain stem and the cerebellum. The cranial nerves emerging from the brain stem show a characteristic arrangement and therefore are easy to define. The arteries of the posterior fossa, however, are more variable, and confusion about the nomenclature of the branches of the VAs and BA may therefore arise. Symmetry – as is always the case with the cranial nerves – can be considered an exception rather than the rule. Usually, three major pairs of arteries supplying the cerebellum and part of the brain stem are distinguished:

1. *The Posterior Inferior Cerebellar Arteries (PICAs).* These vessels originate from the VAs and supply the posterior inferior surface of the cerebellar hemi-sphere and part of the brain stem. There is a wide variation in the course of the PICA and its area of supply. According to Stopford (1916a, b), who studied 150 specimens, the PICA followed a "normal" course in only 58% of cases on the right side and in only 49% of cases on the left. When following this "normal" course, the artery curves around the lower border of the olive and passes in a dorsal direction within the fissure between the medulla oblongata and the cerebellar tonsil, where it forms a caudal loop. In some cases, this loop, which is often referred to as the tonsillar loop, reaches as far as the foramen magnum. Distal to the loop, the artery usually divides into a medial and lateral branch. The medial branch courses in a rostral direction until it reaches the choroid plexus of the fourth ventricle. It then turns in a medial

3

and, subsequently, in a caudal and dorsal direction toward the fissure between the cerebellar tonsil and the uvula and supplies the basal part of the cerebellar vermis. A lateral branch supplies the inferior surface of the cerebellar hemisphere.

However, as has been mentioned above, there is considerable variation in the course and the distribution of the artery. According to Stopford (1916a, b), in 15% of the cases examined by him the artery was absent on the right side, and in 6% of the cases it was absent on the left. Further, an abnormal origin, from the proximal part of the BA, was noted in three cases. Tschernyscheff and Grigorowsky (1929) found this abnormal origin in 15 of 54 cases. Greitz and Sjögren (1963) reported that the artery was absent in two cases. They divided their material into three groups: In the first group, the PICA arises from the VA and approaches the posterior pole of the tonsil; in the second group, the artery has a similar origin but approaches the anterior pole of the tonsil; in the third group, the artery originates from the BA and reaches the anterior pole of the tonsil. Four PICA branch types were distinguished by Stopford (1916a, b):

Bulbar branches

A choroidal branch to the plexus of the fourth ventricle

A posterior spinal branch

Cerebellar branches (medial and lateral), which anastomose freely with branches of the anterior inferior and superior cerebellar arteries

2. The Anterior Inferior Cerebellar Arteries (AICAs). These arteries usually arise from the BA and course in a laterocaudal direction toward the cerebellopontine angle (CPA) and the great horizontal fissure. The details of the course and the area of distribution of these arteries will be described in Sect. 2.1.2.

3. The Superior Cerebellar Arteries (SCAs). These vessels also arise from the BA just before its bifurcation into right and left posterior cerebral arteries. The course of these arteries is more constant than that of the PICAs and AICAs. The artery curves around the pons, immediately caudal to the oculomotor nerve. According to Lazorthes (1961), this artery has a double origin in 10% of cases. Stopford found duplication of this vessel in 12% of cases on the right, in 16% on the left, and in 3% on both right and left sides. The site of origin is almost always just proximal to the bifurcation of the BA into the posterior cerebral arteries. In 6% of Stopford's material the origin was more caudal, whereas in only one case the SCA arose from the junction of the middle and upper thirds of the BA. Absence of the artery was noted only once by Stopford.

Blackburn (1907) examined 220 cases and found the artery was always present.

Three types of branches were distinguished by Stopford:

Pontine branches

Mesencephalic branches, which had already been described by Alezaïs and d'Astros in 1892

Cerebellar branches, which supply mainly the superior surface of the cerebellar hemisphere

Krayenbühl and Yasargil (1957) described a lateral and a medial branch of the SCA. The lateral branch, which is referred to as the marginal artery

4

(Critchley and Schuster 1933), curves toward the frontal edge of the cerebellum and supplies parts of the quadrangular and the superior semilunar lobules. The medial branches of the right and left sides approach each other in the midline and supply the upper part of the vermis.

As mentioned above, terminal branches of the SCAs anastomose with branches of the PICAs and the AICAs of both the ipsilateral and the controlateral side. Krayenbühl and Yasargil (1957) termed this vascular network the "hemispherical plexus."

Apart from the AICAs and the SCAs, the BA gives rise to a large number of small arteries. Stopford (1916a, b) divided these branches into a median set and a set of transverse rami. The median set is composed of minute branches which arise from the side of the BA in contact with the brain stem. These small arteries immediately enter the pons. The transverse rami extend laterally and are often arranged symmetrically. Their number is variable. Stopford also mentioned a larger branch that extends to the trigeminal nerve.

Foix et al. (1925) divided the transverse rami into a group of short circumferential and long circumferential arteries. The same subdivision was made by Böhne (1927). Walker (1965) pointed out the confusion that exists as regards the nomenclature of some of these arteries. In many textbooks of anatomy, the internal auditory artery is described as a branch of the BA. More recent investigations (e.g., Martin et al. 1980) have shown that the internal auditory artery usually arises from the AICA, not from the BA.

2.1.2 Detailed Description of the Anterior Inferior Cerebellar Artery

It must first be stated that there is much variation in the course and size of the AICA. Moreover, there is no relation between the anatomy of this vessel on the right and that on the left side; asymmetry can be considered the rule.

2.1.2.1 Origin

The AICA arises from the BA. Some authors claim that this is always the case (Atkinson 1949; Lazorthes et al. 1950a, b; Maillot et al. 1976). Others note that, exceptionally, the AICA originates from the VA, especially when the PICA is absent on the same side (Goodhart and Davison 1936; Mitterwallner 1955; Krayenbühl and Yasargil 1957). In 1.25% of the cases investigated by Krayenbühl and Yasargil, the AICA originated from the PICA. Salamon and Huang (1976) stated that the AICA is a branch of the PICA in 1% of cases. Gerald et al. (1973) mentioned that in one case the right AICA arose from the left PICA. Scotti (1975) and Ito et al. (1980) reported cases in which an anomalous branch of the carotid artery had a distribution that was nearly equal to that of the AICA. Many of these different findings can be explained by the use of different nomenclatures. It depends on whether an artery is termed after its origin or after its area of distribution. If the latter is the case, an artery with the usual course and area of distribution of the AICA may be so named, even when it arises from the VA or from the PICA. However, if an artery is only named AICA when it arises from the BA, then, as in the case described above in which the artery originates from the PICA, the AICA must be considered to be absent and the PICA as having a larger area of distribution than usual. Lazorthes et al. (1950a, b) distinguished four types of AICAs according to their areas of distribution. His type IV AICA supplies the total inferior aspect of the cerebellar hemisphere, and in this case no PICA exists. This type may also be interpreted as a variation, with the AICA and PICA arising as a common trunk from the BA. Blackburn (1907) observed that the AICA gives rise to a large branch which takes over the role of the PICA when the latter vessel is small or absent.

Stopford (1916a, b) found that the AICA originates in 78% of cases from the lower third of the BA. Other authors report that the AICA originates in most cases from the lower third of the BA (Goodhart and Davison 1936; Krayenbühl and Yasargil 1957; Salamon and Huang 1976) or from the proximal half of the BA (Gerald et al. 1973). Lazorthes and co-workers (1950a, b, 1976), however, found that in the majority of cases the artery arises from the middle third of the BA.

From these different findings we may conclude that there is a wide variation in the site of origin of the AICA. This is in accordance with the findings of Naidich et al. (1976a, b). The site of origin in their material was the lower or the middle third of the BA in 53% and 47% of cases, respectively.

2.1.2.2 Multiplication of the AICA

Blackburn (1907) investigated 220 cases and found duplication of the AICA in only eight. Stopford (1916a, b) once found a double AICA, and Mitterwallner (1955), who examined 360 specimens, saw no more than 17 cases in which the AICA was double.

Much higher percentages were found by Martin et al. (1980): A single AICA was seen in 72%, duplicate arteries in 26%, and triplicate arteries in 2% of cases. Salamon and Huang (1976) found about the same: A solitary artery was found in 73%, paired arteries in 25%, and triplicate arteries in 1% of cases.

Naidich et al. (1976a, b) found that, in 29% of the specimens investigated, there was a duplication of the AICA. The larger of the two vessels was referred to as the AICA and the smaller as an accessory artery. They distinguished between an inferior and a superior accessory artery, depending on the site of origin from the BA – caudal or rostral to the AICA. In 10% of cases, they found two "aberrant" arteries, neither of which showed the typical course of the AICA. A similar finding was mentioned by Gillilan (1964), who found the AICA infrequently replaced by several smaller vessels originating from the basilar trunk. Ferrari Lelli (1939), in a study of the internal auditory artery, described one case in which a small artery, originating from the BA, gave off branches to the inner ear as well as to the cerebellum. He called this vessel "arteria cerebelli accessoria." Lazorthes et al. (1950a, b) mentions some cases in which the internal auditory artery arises from the BA just below the origin of the AICA. This artery was referred to as "artère supplémentaire moyenne basse." This finding has been confirmed by others (Fisch 1969; Mazzoni 1969).

Jakob (1928) described a so-called middle inferior cerebellar artery which is also mentioned by Tschabitscher and Perneczky (1974). This artery is situated between the AICA and SCA. Lazorthes et al. (1950a, b) proposed for this vessel the name "artère supplémentaire moyenne haute." However, in most publications this artery is not mentioned and is probably regarded as a (superior) accessory artery.

2.1.2.3 Absence of the AICA

Absence of the AICA is a rare phenomenon. Tschabitscher and Perneczky (1974) found the artery always present in the 44 brains they examined. In the 220 cases examined by Blackburn (1907), the artery was absent in only seven cases. In Stopford's material (1916a, b), the AICA was missing once on one side and once on both sides. Mitterwallner (1955) reported absence of the AICA in 2.7% and Salamon and Huang (1976) in 2% of cases. Lazorthes et al. (1950a, b) found the artery missing in 2 of 30 cases. Krayenbühl and Yasargil (1957), however, found much higher figures: The AICA was absent in 7.5% of cases on the right, in 10.5% on the left, and in 2% on both sides.

It is unclear how the differences between the figures of Krayenbühl and Yasargil and those of other authors can be explained. Possibly, a small vessel originating from the BA, which runs toward the internal auditory meatus and gives off only small branches to the cerebellum, is not considered an AICA by Krayenbühl and Yasargil, whereas other authors regard such a small vessel as a variant of the AICA.

2.1.2.4 The Course of the AICA in Relation to the Abducens Nerve

After its origin from the BA, the AICA usually turns in a lateral or caudolateral direction toward the flocculus, often with a superior concave curve. The artery thus crosses the abducens nerve. Cushing (1910) was the first to study in detail the relation between the lateral branches of the BA and the abducens nerve. The reason for this study was his observation that in cases of increased intracranial pressure, abducens palsies often occur and that such palsies may be caused by vascular strangulation of the nerve. In the majority of his cases, the arteries were more or less symmetrically disposed and overlay the abducens nerves in such a way as to make strangulation possible. In other cases, the arteries crossed so far posteriorly that constriction of the nerves seemed unlikely. In 10 of 59 cases examined, either the right or the left nerve was superficial to the artery, i.e., the artery passed between the abducens nerve and the pons.

The same observations were made by Stopford (1916a, b), Konaschko (1927), Watt and McKillop (1935), Lazorthes et al. (1950a, b), and Krayenbühl and Yasargil (1957).

Perforation of the abducens nerve by the AICA can be considered rare (Cushing 1910; Mitterwallner 1955). Stopford (1916a, b) described eight cases in which the artery on one side was too far forward to have any relation to the nerve. He further pointed out that strangulation of the abducens nerve in the case of increased intracranial pressure occurs more frequently when the artery is in a dorsal position to the nerve.

Naidich et al. (1976a) made the interesting observation that the relation of the AICA to the abducens nerve depends on the distance between the pontomedullary sulcus and the point at which the artery crosses the nerve. When the crossing is within 6 mm of the pontomedullary sulcus, the artery is situated ventral to the nerve. When the crossing is 6–8 mm from the pontomedullary sulcus, either a ventral or a dorsal position may occur. When the crossing is 8–11 mm from the sulcus, the artery is always situated dorsal to the nerve. Finally, when the AICA is 11 mm or more rostral to the sulcus, it bears no relation to the nerve.

2.1.2.5 Relation of the AICA to the Facial and Vestibulocochlear Nerves and to the Internal Auditory Meatus

After crossing the abducens nerve, the AICA passes toward the cerebellopontine angle. Here, it often bears a close relation to the facial and vestibulocochlear nerves and to the internal auditory meatus. In most instances, the artery forms a loop in this region According to Lazorthes et al. (1950a, b), the concavity of the loop is directed inward and backward. This loop may be called cerebellar (Mazzoni 1969; Mazzoni and Hansen 1970) or meatal (Naidich et al. 1976a, b). Naidich and co-workers distinguished between a single and a double arterial loop. The double loop consists of a meatal and a brachial loop. The latter is located on, or close to, the brachium pontis. Tschabitscher and Perneczky (1974) examined 85 cerebellopontine angles: In 14 cases, no meatal loop was seen; in 13 cases, the loop was within the meatus; in 16 cases, it just reached the internal auditory orifice; in the remaining 41 cases, the loop did not reach as far as the orifice.

Usually, the loop has less curvature when it just reaches the orifice and does not enter the meatus (Mazzoni 1969, 1970). Mazzoni pointed out the difficulty in expressing the relation between the artery and the nerves with the terms "anterior," "inferior," etc., because the artery often has a tortuous course. In 35 of his 67 cases, the loop of the artery passed between the facial and the vestibulocochlear nerves. In the majority of the other 32 cases, the AICA passed inferior or anterior to the nerves. Similar observations were made by Martin et al. (1980) in their extensive study of the anatomic relations between the artery and the facial and vestibulocochlear nerves. Sunderland (1945, 1948) distinguished between no fewer than ten different types of artery-nerve relation.

Tschabitscher and Perneczky (1974), on the other hand, described only three types – internerval, ventrobasal, and dorsal types. Watt and McKillop (1935) made a similar classification. Apart from a dorsal, a ventral, and two internerval positions, these authors described a situation in which the loop runs around the nerves.

2.1.2.6 Branches of the AICA

From the proximal part of the AICA, small branches arise which are distributed to the lateral aspect of the inferior pontine tegmentum (Böhne 1927; Gillilan 1969). Some branches may extend on the surface of the pons as far caudally as the region of the rostral medulla, before penetrating the brain substance (Kaplan 1959). Most authors describe two main branches of the AICA, a lateral and a medial branch, which arise within the cerebellopontine angle cistern (Atkinson 1949; Gerald et al., 1973; Perneczky and Tschabitscher 1976; Ross and Du Boulay 1976; Takahashi 1968, 1974). This bifurcation of the AICA occurs just before or after the point at which the artery crosses the facial and vestibulocochlear nerves. The lateral branch courses around the upper edge of the flocculus toward the great horizontal fissure. On the cerebellar hemisphere, it anastomoses with branches of both the PICA and the SCA. The medial branch courses in a caudomedial direction toward the brain stem and supplies the inferomedial part of the cerebellar hemisphere. It usually anastomoses with branches of the PICA. Martin et al. (1980) described a similar anatomic relation, but the lateral and medial branches are defined by them as rostral and caudal trunks.

The most comprehensive description of the AICA branches is given by Naidich et al. (1976a, b). This differs in some respects from the description given by other authors. Naidich et al. also distinguish two main branches of the AICA, which are termed rostrolateral and caudomedial branches. However, according to these authors, the bifurcation occurs at the crossing with the abducens nerve or a few millimeters lateral to it. The rostrolateral branch (RLB) or rostrolateral artery forms the meatal loop. Distal to the loop, the artery enters the supraflocular portion of the great horizontal fissure. Near the posterosuperior corner of the flocculus, the artery divides into an ascending branch toward the horizontal fissure and a descending branch toward the posterolateral fissure. The ascending branch of the rostrolateral artery often reaches the occipital aspect of the cerebellum. The descending branch courses in the posterolateral fissure between the flocculus anteriorly and the biventral lobule posteriorly. From this so-called retroflocular segment small branches pass to the inferior semilunar and biventral lobules. Naidich et al. point out the relation which exists between this descending branch and the PICA. When the latter artery is small, there is often a large descending branch, which runs on the lateral surface of the biventral lobule or in the cerebellomedullary fissure toward the posterior inferior aspect of the cerebellar hemisphere. According to Naidich et al., the caudomedial branch (CMB) is consistently found, although it is quite variable in size. It arises in the vicinity of the abducens nerve and describes a caudal loop on the lateral aspect of the pons and the medulla oblongata. Sometimes it is only a tiny branch which terminates on the pons. When the CMB is larger, the artery also supplies the lateral aspect of the flocculus and the biventral lobule. In one-third of the cases investigated by Naidich et al. the CMB is quite large and forms a laterally directed loop around the roots of the ninth, tenth, and eleventh nerves, the foramen of Luschka, and the "choroid plexus of the fourth ventricle." Small branches to the choroid plexus of the lateral recess of the fourth ventricle arise from this part of the CMB (Maillot et al. 1976; Fujii et al. 1980).

Distal to the loop described above, the CMB continues its course in a way similar to that of the descending branch of the rostrolateral artery, and it may replace the PICA or a part of it. The CMB is usually smaller than the RLB. Only when the CMB replaces the PICA, or when the RLB is replaced by the marginal branch of the SCA, is the reverse seen. Naidich et al. emphasize that different names for the branches of the AICA have been used by various authors. The bifurcation of the AICA is usually considered to be located in the cerebellopontine angle, not in the vicinity of the abducens nerve. According to Naidich et al., however, this is not the bifurcation of the AICA itself, but the bifurcation of the RLB into its ascending and descending branches. Salamon and Huang (1976) distinguished between three AICA branches where the PICA is absent – an external, a middle, and an internal hemispheric branch. Although the exact course and site of origin of these branches are not specified by them, one may assume that these are the same three arteries that are described by Naidich et al.: the ascending branch of the RLB, the descending branch of the RLB, and the CMB, respectively.

Apart from these two or three main branches, the AICA gives of many smaller branches. The branches to the pons and medulla oblongata and the choroid branches have already been mentioned.

Perneczky and Tschabitscher (1976) divided the AICA into four topographical segments. The so-called anterior pontine segment, which is the most proximal part of the artery and measures about 1 cm, rarely gives off any branches. From the lateral pontine segment the pontine branches arise. The next segment is the cerebellopontine angle segment. From this part of the artery several branches originate, of which the "labyrinthine artery," the subarcuate artery, and the arteries which accompany the facial and vestibulocochlear nerves are the most important. Much attention has been paid to the small arteries in the cerebellopontine angle region. This is due to the important role these branches play in neurosurgical procedures in the cerebellopontine angle. The anatomy of the internal auditory artery (IAA) has been particularly extensively studied (Nabeya 1923; Konaschko 1927; Ferrari Lelli 1939; Sunderland 1945; Nager 1954; Charachon and Latarjet 1962; Fisch 1969; Mazzoni 1969, 1970; Smaltino et al. 1971).

Cavatorti (1908) found that in 70% of cases the IAA arises directly from the basilar trunk. Stopford (1916a, b) found that the IAA originates from the BA in quite a high percentage of cases, namely in 36% on the right and in 28% on the left side. This contrasts with the findings of Ferrari Lelli (1939), Adachi (1928), and Guerrier and Villacèque (1949), who never observed the IAA originating from the BA. Watt and McKillop (1935) and Lazorthes et al. (1950a, b) regarded a separate origin of the IAA from the BA as an exception. Smaltino et al. (1971) described a so-called cerebellolabyrinthine artery. In 87% of cases, this artery originated from the AICA and in 10% directly from the basilar trunk. In the remaining 3%, the artery arose as a separate branch from the BA. From Smaltino's description it is clear that the cerebellolabyrinthine artery, when it originates from the AICA, is the same as the rostrolateral branch of the AICA, described by Naidich et al. (1976a, b). Konaschko (1927) did not distinguish an AICA at all; in its entire length, the artery was named the internal auditory artery. Mazzoni (1969) described multiplication of the IAA. He found a single IAA in 51 of 100 cases, a double IAA in 45 cases, and triplication of the artery in four cases. He also mentioned three cases in which the IAA did not arise from the AICA or from an accessory AICA, but from the PICA.

A comprehensive study of the arteries in the cerebellopontine angle has recently been published by Martin et al. (1980). These authors report that IAAs always originate from a branch of the AICA. Every artery that enters the internal auditory canal is regarded by them as an IAA. Having defined the artery in this way, multiplication was observed in as many as 70% of cases. In 2%, four IAAs were seen. Of all IAAs, 77% arose from the premeatal segment of the AICA. Apart from the IAA, in 82% of the cerebellopontine angles investigated, recurrent perforating arteries were found. The latter arteries arise from nerve-related vessels and often run first toward the meatus before taking a recurrent course to the brain stem. They travel along the facial and vestibulo-cochlear nerves and give off branches to these nerves. In the majority of cases, these arteries are single. A subarcuate artery was present in 72% of the cases investigated. This is an artery which penetrates the dura, enters the subarcuate canal and supplies the mastoid in the region of the semicircular canals. In only 4 of 50 cerebellopontine angles investigated was a so-called cerebellosubar-cuate artery seen. This artery sends one branch to the subarcuate fossa and one branch to the cerebellum. Martin et al. explained the different findings of the various authors with regard to the origin of the IAA by differences in definitions of the AICA and the IAA. In accordance with the studies of Adachi (1928) and Fisch (1969), they call an artery which arises from the BA an AICA when it gives off branches to the cerebellum. The site of origin of the IAA is defined by them as the point at which the branch to the internal auditory canal originates.

2.1.2.7 Variations of the AICA

From the previous description it is clear that many variations of the AICA exist. The artery may be absent, or it may be extremely well developed, especially when on the same side the PICA is absent. It may be duplicated or even tripli-cated, and it may arise either from the VA or from the BA. The various branches of the AICA may differ in length. The characteristic meatal loop is sometimes absent but, if present, may be either part of the main trunk of the AICA or part of the RLB. Takahashi et al. (1968) stated that variations of both the PICA and the AICA are the rule rather than the exception. He also found that the size of the AICA is usually inversely related to the size of the PICA. This pattern is recognized by most other authors.

According to the territory supplied, Lazorthes et al. (1950a, b, 1961) distin-guished four types of the AICA. In the first type, the artery supplies only the flocculus. In the second type, the AICA extends to the anterior aspect of the cerebellum up to the lateral border. The third type occurs when the artery additionally supplies the anteroexternal part of the inferior aspect of the cerebellar hemisphere. In the fourth type, the AICA supplies both the total anterior and the total inferior aspect of the cerebellar hemisphere. In this latter case, the PICA is either small or absent.

2.1.2.8 Areas of Supply

It is obvious that the areas of supply differ with the length and the course of the AICA. As many variations in the length of the vessel exist, it is clear that similar variations are observed in the areas supplied by the AICA.

The lateral and caudal parts of the pons are supplied by branches of the AICA (Stopford 1916a, b; Böhne 1927; Atkinson 1949; Hiller 1952; Krayenbühl and Yasargil 1957; Kaplan 1956, 1959; Gillilan 1964; Brouckaert et al. 1981; Perneczky et al. 1981). The number of these branches is proportional to the caliber of the artery from which they arise (Gillilan 1964). Structures that are supplied by branches of the AICA are the facial nucleus, the spinal nucleus and spinal tract of the trigeminal nerve, and some or all of the fiber bundles of the ascending lemniscal system (Gillilan 1969). Together with the PICA, the AICA supplies a variable part of the upper medulla oblongata (Stopford 1916a, b; Krayenbühl and Yasargil 1957; Fisher et al. 1961; Gillilan 1964). The middle cerebellar peduncle or part of it may also be supplied by the AICA (Atkinson 1949; Tschernyscheff and Grigorowski 1929). The AICA further supplies a variable part of the cerebellar hemisphere.

According to Fazzari (1929), who investigated the AICA and other cerebellar arteries in several animals, the flocculus and paraflocculus are always supplied by branches of the AICA. Lazorthes et al. (1950a, b) pointed out that the AICA supplies the phylogenetically oldest part of the cerebellum (the archicerebellum), i.e., the flocculus and nodulus. This part of the cerebellum is closely related to the vestibular apparatus, which is supplied by the IAA. For this reason, the term "cerebellolabyrinthine system," used by Guerrier and Villacèque (1949) to define the complex of AICA and IAA, was regarded by Lazorthes et al. as an excellent term.

Depending on the length of the AICA, it may supply only the flocculus and paraflocculus, or it may supply other parts of the inferior aspect of the cerebellar hemisphere as well. According to Tschernyscheff and Grigorowski (1929), parts of the anterior quadrangular lobule, the posterior quadrangular lobule, the biventral lobule, and the cerebellar tonsil may be supplied by the AICA. Parts of the gracilis lobule, the inferior semilunar lobule, and the superior semilunar lobule are in some cases also vascular territories of the AICA.

The AICA does not contribute to the supply of any of the deep cerebellar nuclei (Gillilan 1969).

2.1.2.9 Anastomoses Between the AICA and Other Cerebellar Arteries

Several authors describe anastomoses between the hemispherical branches of the AICA, the PICA, and the SCA (Goodhart and Davison 1936; Atkinson 1949; Krayenbühl and Yasargil 1957; Mani et al. 1968; Ross and DuBoulay 1976; Takahashi 1974). According to Tschabitscher et al. (1975), these anastomoses mainly occur in: (a) the roof of the fourth ventricle; (b) the paravermian region; (c) the horizontal fissure; and (d) the tentorial surface of the cerebellar hemisphere. The CMB of the AICA has anastomoses with branches of the PICA, whereas the RLB anastomoses with branches of both the PICA and the SCA.

2.2 Injection Study of the Vertebrobasilar Arteries

2.2.1 Materials and Methods

The present investigation was carried out on the brains of 20 patients, who, in a majority of cases, died of cardiac disease and did not suffer from any neurologic disease during life. Ages varied from 51 to 79 years.

The brains were carefully removed in toto within 24 h of death. The brain stem and VAs were cut as far caudally as possible in the vicinity of the foramen magnum. The right and left posterior cerebral arteries were ligated immediately distal to their origins from the BA. Catheters were inserted in both VAs and the arterial system was perfused with a warm heparin solution. Subsequently, a heated, red-colored solution of gelatine and barium sulfate was injected. The specimens were then cooled in order to allow the gelatin to set. Fixation took place in 10% formaline over a period of 1–4 months.

The frontobasal aspect of the brain stem and the cerebellum of a normal brain are illustrated in Fig. 1, and the course of the arteries of each brain is indicated. In addition, multidirectional roentgenograms of the preparations were made.

Because the examination of the arteries could be performed only after removal of the brains from the skulls, the exact relations between arteries and nerves in the cerebellopontine angle could not be studied. For this reason, the IAA will not be discussed in the following paragraph.

With the following exceptions, the nomenclature of Naidich et al. (1976a, b) is used:

The ascending and descending branches of the RLB, which are named by Naidich et al., will not be named separately.

The concept of a common trunk of the AICA and the PICA is not used.

An artery arising from the BA and supplying the territories usually supplied by the AICA and the PICA is termed AICA. On the other hand, an artery arising from the VA and supplying both territories is referred to as the PICA.

When a PICA with its usual area of supply originates from the VA, and there is another artery which supplies the usual AICA territory but arises also from the VA, the latter is called AICA despite its unusual origin. A second artery that supplements or replaces a part of the area of supply of the AICA and arises from the BA is termed either a superior or an inferior accessory artery. This is in accordance with the nomenclature of Naidich et al.

2.2.2 Results

In case 1 (Fig. 2), the VAs join each other caudal to the pontomedullary fissure. On both sides, a PICA arises from the VA. On the right side, the AICA originates from the middle of the BA. The artery crosses the abducens nerve on the ventral side. When the artery reaches the facial nerve, it divides into a rostrolateral (RLB) and a caudomedial branch (CMB). The RLB follows the facial and vestibulocochlear nerves and forms a complicated meatal loop (ML). It follows its course through the horizontal fissure (HF) in a lateral direction

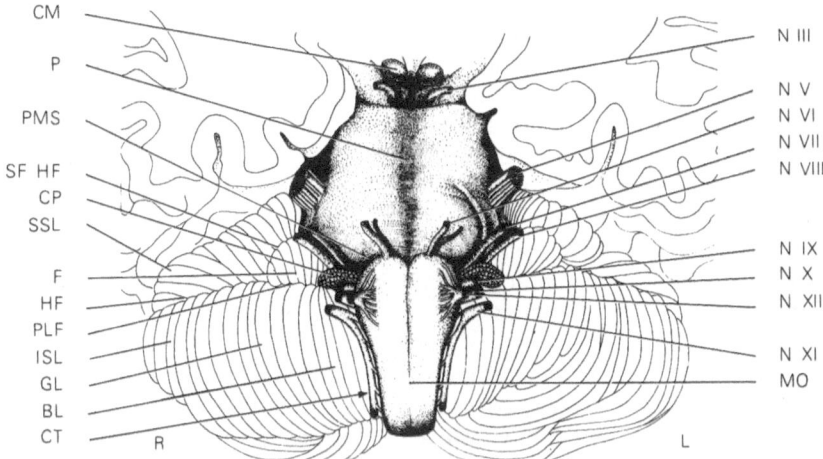

CM   N III
P   N V
PMS   N VI
SF HF   N VII
CP   N VIII
SSL
F   N IX
HF   N X
PLF   N XII
ISL   N XI
GL   MO
BL
CT   R   L

Fig. 1

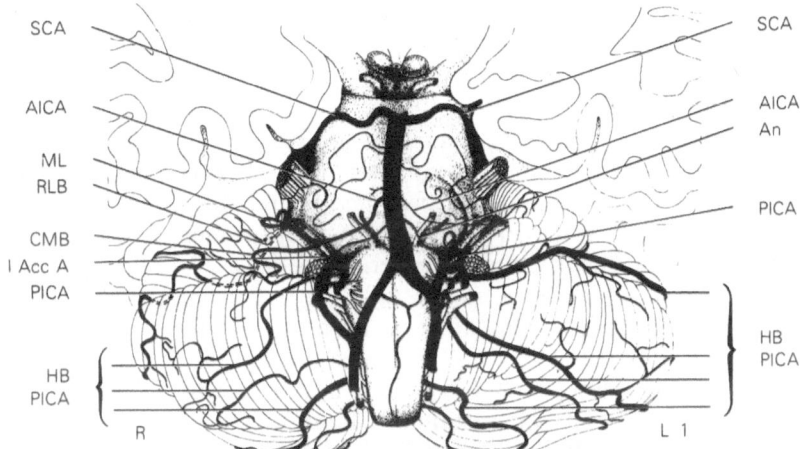

SCA   SCA
AICA   AICA
           An
ML
RLB   PICA
CMB
I Acc A
PICA   HB
           PICA
HB
PICA
   R   L 1

Fig. 2

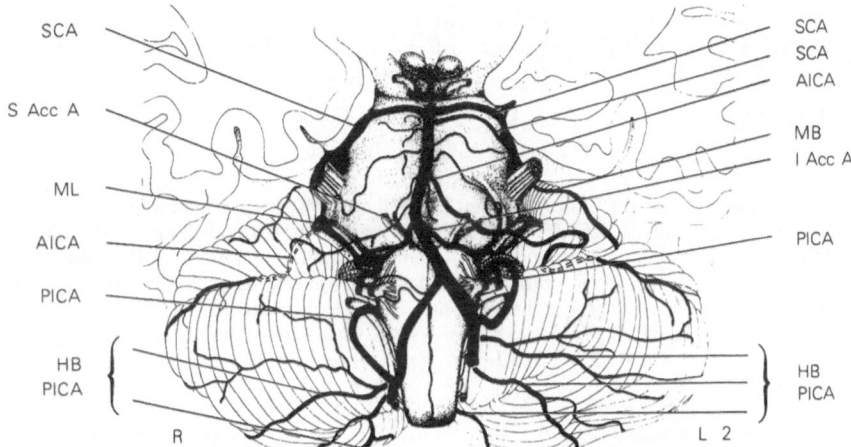

SCA   SCA
           SCA
           AICA
S Acc A
           MB
           I Acc A
ML
AICA   PICA
PICA
HB   HB
PICA   PICA
   R   L 2

Fig. 3

14

and divides into branches which anastomose with the hemispherical branches (HB) of the PICA on the lateral border of the cerebellar hemisphere. The CMB passes ventral to the flocculus and turns in an opposite direction near the suprafloccular part of the HF. Through the PLF, the artery continues its course toward the lateral recess of the fourth ventricle.

There is a small inferior accessory AICA (IAccA) on the right side. On the left side, a small AICA is present, which extends laterally as far as the flocculus. Caudal to the AICA, a small vessel forms an anastomosis (An) between the BA and the PICA. One of the hemispherical branches of the PICA (HB) courses in the HF where normally the AICA runs. The right and left SCA are symmetrical.

In case 2 (Fig. 3), the VAs join each other caudal to the pontomedullary fissure. On the right side, the PICA arises from the caudal part of the VA. The posterior inferior aspect of the cerebellar hemisphere is supplied by its branches (HB PICA). The AICA arises from the unusually long BA at about the level of the junction of the lower and middle thirds. It crosses the abducens nerve ventrally and forms a loop near the origin of the facial and vestibulocochlear nerves. It then follows the nerves in the "internerval position" up to the internal auditory meatus, where the meatal loop (ML) is formed. The AICA continues its course in the HF.

Distal to the origin of the AICA, a small superior accessory artery (SAccA) arises from the BA. This artery also crosses the abducens nerve ventrally and then crosses the facial and vestibulocochlear nerves and the AICA ventrally. It supplies the flocculus. On the left side, the PICA arises more rostrally than on the right side. Its branches supply the posterior inferior aspect of the cerebellar hemisphere. At the same level as the right AICA, a small artery arises from the BA, crosses the abducens nerve at its base, and reaches the facial and vestibulocochlear nerves. This vessel is termed the inferior accessory artery (IAccA). Rostral to this artery, the AICA arises. The AICA crosses the abducens nerve dorsally and the facial and vestibulocochlear nerves ventrally. On the surface of the flocculus, a loop is formed, after which the AICA continues its course in the posterolateral (PLF) and the horizontal fissure (HF). The SCA is a single artery on the right side and is duplicated on the left side.

In case 3 (Fig. 4), the VAs unite near the pontomedullary sulcus. The BA has a curved course with the convexity of the curve to the right side. No PICA is present on the right side. The right AICA bifurcates just after its ventral crossing with the abducens nerve. The rostrolateral branch (RLB) passes lateral-

Fig. 1. Frontobasal view of the brain stem and cerebellum. *CM*, corpus mammillare. *P*, pons. *PMS*, pontomedullary sulcus. *SF HF*, suprafloccular part of horizontal fissure. *CP*, choroid plexus of lateral recess of fourth ventricle. *SSL*, superior semilunar lobule. *F*, flocculus. *HF*, horizontal fissure. *PLF*, posterolateral fissure. *ISL*, inferior semilunar lobule. *GL*, gracilis lobule. *BL*, biventral lobule. *CT*, cerebellar tonsil. *NIII*, oculomotor nerve. *NV*, trigeminal nerve. *NVI*, abducens nerve. *NVII*, facial nerve. *NVIII*, vestibulocochlear nerve. *NIX*, glossopharyngeal nerve. *NX*, vagal nerve. *NXI*, accessory nerve. *NXII*, hypoglossal nerve. *MO*, medulla oblongata

Fig. 2. Case 1. *An*, anastomosis

Fig. 3. Case 2

Fig. 4. Case 3

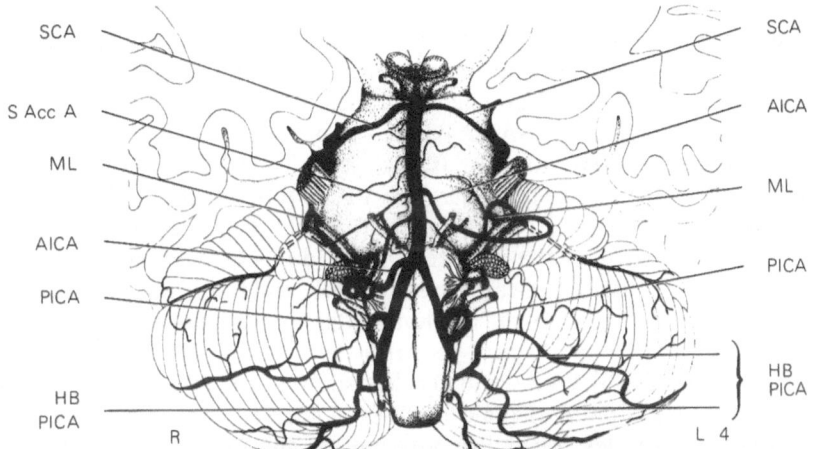

Fig. 5. Case 4

Fig. 6. Case 5

16

ly between the facial and vestibulocochlear nerves and does not form a loop. It runs partially within the horizontal fissure and gives off branches to both the biventral and the superior semilunar lobules. The caudomedial branch (CMB) runs in a caudal direction in the cerebellomedullary fissure. Its branches (HB CMB) supply the posterior inferior aspect of the right cerebellar hemisphere and anastomose with branches of the RLB.

From the left VA, a PICA arises that has the usual area of supply. The left AICA arises from the BA at about the same level as the right AICA. It crosses the abducens nerve ventrally and the facial and vestibulocochlear nerves dorsally and does not form a meatal loop. No caudomedial branch is observed. A single SCA is present on both sides.

In case 4 (Fig. 5), the VAs join each other near the pontomedullary fissure. The branches of the right PICA (HB PICA) supply the posterior inferior part of the right cerebellar hemisphere. At the level of the junction of both VAs, an artery arises that bears no relation to the pons or to the abducens nerve. If forms a complicated loop in the vicinity of the origin of the glossopharyngeal, the vagal, the accessory, and the hypoglossal nerves and then runs along the facial and vestibulocochlear nerves up to the internal auditory meatus, where the meatal loop (ML) is formed. The artery continues through the horizontal fissure in a lateral direction. No caudomedial branch is present.

A superior accessory artery (SAccA) arises about halfway from the BA and crosses the abducens nerve dorsally. It reaches as far laterally as the facial and vestibulocochlear nerve complex. The left PICA follows about the same course as the right PICA. The left AICA arises from the BA at the same level as the right SAccA. It crosses the abducens nerve dorsally. Subsequently, it crosses the facial and vestibulocochlear nerves ventrally and forms a wide meatal loop (ML). It runs laterally through the horizontal fissure. No caudomedial branch can be observed. The SCAs are symmetrical.

In case 5 (Fig. 6), both the VAs and the BA show a tortuous course. The junction of the VAs takes place at the origin of the left abducens nerve. From the right VA, a strongly developed PICA originates that supplies the major part of the right cerebellar hemisphere. One of its branches reaches the horizontal fissure and anastomoses with the marginal branch (MB).

Only a very small AICA is present. It crosses the abducens nerve ventrally and disappears behind the double loop that is formed by the right PICA. This loop reaches as far as the facial and vestibulocochlear nerve complex.

On the left side, the hemispherical branches of the PICA (HB PICA) supply the posterior inferior aspect of the cerebellar hemisphere. The left AICA arises from the junction of the VAs. It runs parallel to the left VA in its first trajectory, crosses the proximal parts of the facial and vestibulocochlear nerves ventrally, then runs parallel to these nerves for a short distance and describes a loop that curves distal to the nerves. Within the suprafloccular part of the horizontal fissure, the artery divides into a rostrolateral (RLB) and a caudomedial branch (CMB). A small superior accessory artery (SAccA) crosses the abducens nerve ventrally. The SCAs both show a normal configuration.

In case 6 (Fig. 7), the junction of the VAs is caudal to the pontomedullary fissure. The branches of the right PICA (HB PICA) supply the posterior inferior part of the cerebellar hemisphere. A small anastomotic branch (An) connects the right PICA with the right AICA.

17

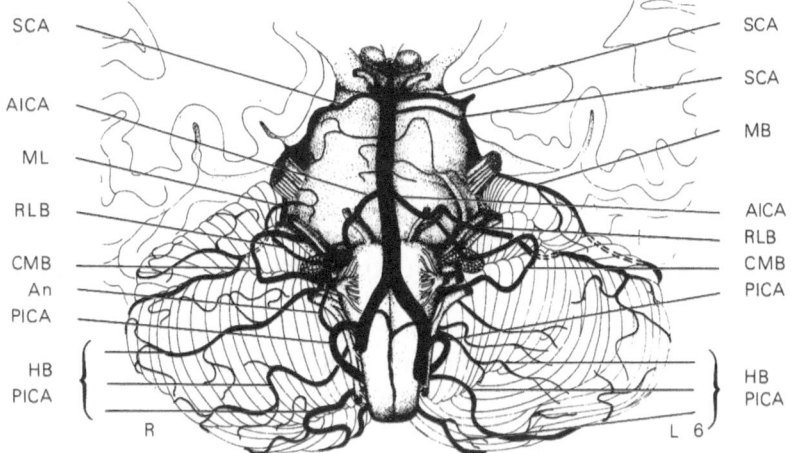

Fig. 7. Case 6. *An*, anastomosis

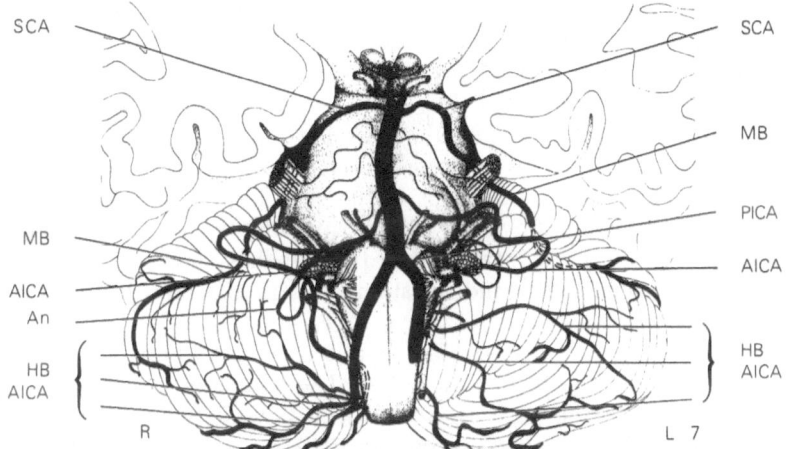

Fig. 8. Case 7. *An*, anastomosis

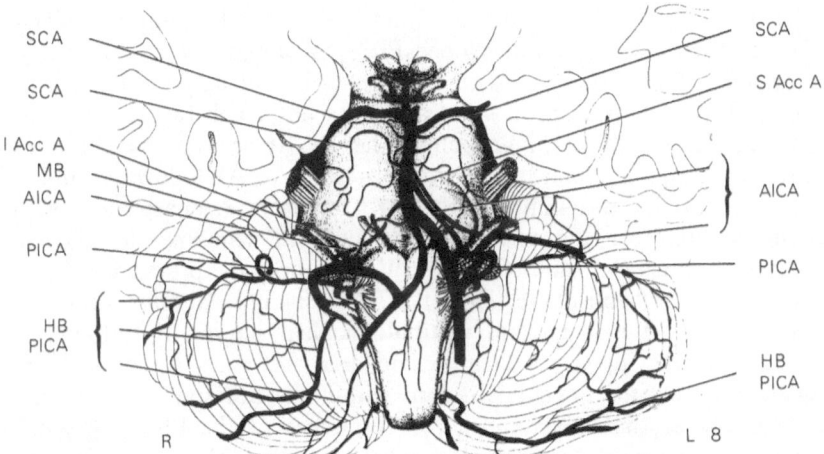

Fig. 9. Case 8

18

The right AICA arises about halfway from the BA and crosses the abducens nerve ventrally. Its main trunk reaches almost as far caudolaterally as the ponto-medullary fissure, where it divides into its two main branches. The caudomedial branch (CMB) passes along the facial and vestibulocochlear nerves to the supra-floccular part of the horizontal fissure and then turns gradually in a medial direction. The rostrolateral branch (RLB) first forms the meatal loop and then passes to the lateral part of the horizontal fissure.

The left PICA is somewhat smaller than the right and supplies a smaller part of the cerebellar hemisphere. The left AICA divides into its two main branches directly after having crossed the abducens nerve. The rostrolateral branch (RLB) crosses the facial and vestibulocochlear nerves ventrally without forming a meatal loop. It traverses the flocculus and runs toward the horizontal fissure. The caudomedial branch (CMB) is somewhat smaller in caliber and crosses the facial and vestibulocochlear nerves more proximally. On the right side, the SCA is a single artery, whereas it is duplicated on the left side.

In case 7 (Fig. 8), the VAs unite near the pontomedullary fissure. On the right side, no PICA is present. The AICA crosses the abducens, facial, and vestibulocochlear nerves ventrally. It traverses the flocculus and describes a wide loop, after which it follows a medially directed course at first, but gradually turns in a more caudal direction and thus traverses the cerebellar tonsil. For a short distance, it runs within the medullocerebellar fissure and then divides into several branches, which supply the posterior inferior aspect of the cerebellar hemisphere. The proximal and distal limbs of the wide loop of the AICA are connected by a small anastomotic vessel (An). The marginal branch (MB) of the right SCA runs in the horizontal fissure. Its branches supply the lateral part of the cerebellar hemisphere and anastomose with the branches of the AICA (HB AICA).

On the left side, there is a comparable situation. A small branch, arising from the left VA, can be described as the PICA. The course of the left AICA is similar to the course of the right AICA. A rostrolateral branch does not exist, and the marginal branch (MB) of the left SCA runs in the lateral part of the horizontal fissure.

The SCA is a single artery on both sides.

In case 8 (Fig. 9), the VAs join each other rostral to the pontomedullary fissure. On the right side, a well-developed PICA is present. One of its hemi-spherical branches (HB PICA) runs in the posterolateral and the horizontal fissure. The right AICA is a short vessel and does not reach further than the complex of the facial and vestibulocochlear nerves. It crosses the abducens nerve ventrally. A smaller inferior accessory artery (AccA) follows a similar course. Branches of the marginal branch (MB) of the SCA anastomose with branches of the PICA on the surface of the superior semilunar lobule. On the left side, the branches of the PICA (HB PICA) supply the posterior inferior aspect of the cerebellar hemisphere. The left AICA runs in a caudolateral direc-tion to the origin of the facial and vestibulocochlear nerves. It follows the course of these nerves for some distance and does not form a meatal loop. After having traversed the flocculus, it runs within the horizontal fissure and divides into branches that anastomose with branches of the PICA. A small superior accessory artery (SAccA) runs parallel to the AICA and reaches the facial and vestibulocochlear nerve complex. One SCA is present on both sides.

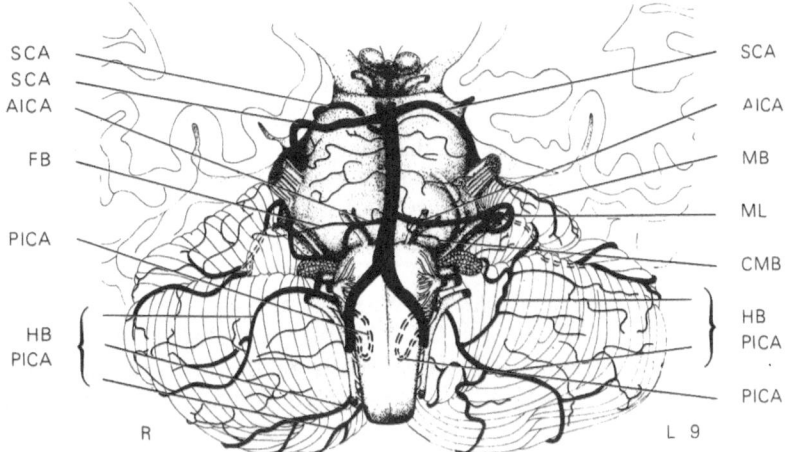

Fig. 10. Case 9. *FB*, floccular branch

Fig. 11. Case 10

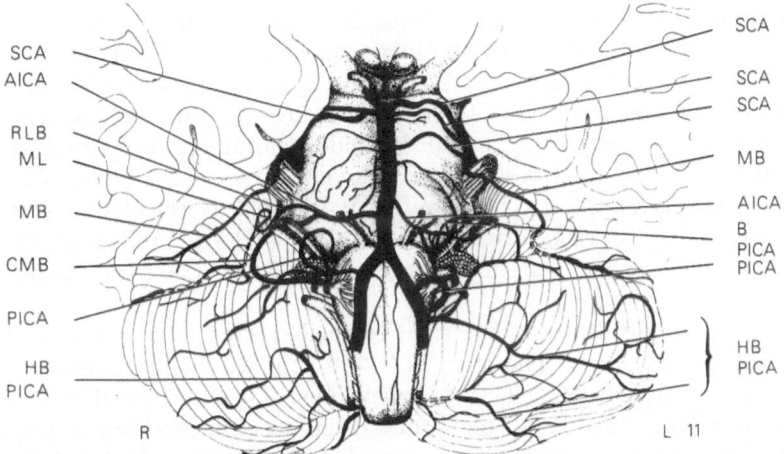

Fig. 12. Case 11. *B PICA*, PICA branch

20

In case 9 (Fig. 10), the VAs unite somewhat caudal to the pontomedullary fissure. On the right side, the branches of the PICA (HB PICA) supply the posterior inferior aspect of the cerebellar hemisphere. The right AICA crosses the abducens nerve ventrally and gives off a small branch (FB) to the flocculus. After the crossing with the abducens nerve, the artery immediately turns caudally and crosses the facial and vestibulocochlear nerves ventrally. On the flocculus, it turns rostrally, passes between the nerves, and turns backward toward the horizontal fissure. Within the suprafloccular part of the horizontal fissure, the artery bifurcates into two branches.

On the left side, one of the hemispherical branches of the PICA (HB PICA) supplies a part of the flocculus. The left AICA passes the abducens nerve ventrally. A tiny caudomedial branch (CMB) arises from the proximal part of the AICA. The meatal loop (ML) lies in close relation with the facial and vestibulocochlear nerves. On the right side a double SCA is present, whereas on the left side the SCA is a single artery. The marginal branch on the left side is more pronounced than on the right side.

In case 10 (Fig. 11), the VAs join each other caudal to the pontomedullary fissure. On the right side, the posterior inferior aspect of the cerebellar hemisphere is supplied by a hemispherical branch of the PICA (HB PICA). The right AICA passes ventral to the abducens nerve. In the pontomedullary fissure, the AICA passes behind the choroid plexus of the fourth ventricle, where it divides into two branches to the anterior inferior aspect of the cerebellar hemisphere. The right AICA does not bear any relation to the internal auditory meatus. The marginal branch (MB) of the SCA supplies the lateral part of the cerebellar hemisphere. Its branches anastomose with branches of both the PICA and the AICA. A small superior accessory AICA (SAccA) is present.

On the left side, the major part of the cerebellar hemisphere is supplied by branches of the PICA (HB PICA). The AICA on this side has a small caliber. It crosses the abducens, the facial, and the vestibulocochlear nerves ventrally and does not bear any relation to the internal auditory meatus. On the surface of the gracilis lobule, the AICA describes a loop and turns backward to the choroid plexus of the fourth ventricle. The marginal branch (MB) of the SCA on both the right and and the left side is well developed.

In case 11 (Fig. 12), the VAs join each other near the pontomedullary fissure. The right PICA is small. The right AICA bifurcates into a caudomedial branch (CMB) and a rostrolateral branch (RLB) between the crossing with the abducens nerve and the crossing with the facial and vestibulocochlear nerves. The CMB passes dorsal to the facial and vestibulocochlear nerves and describes a wide loop along the border of the flocculus toward the medial aspect of the right cerebellar hemisphere. The RLB forms the meatal loop (ML) and runs laterally in the horizontal fissure. The marginal branch (MB) of the SCA runs parallel to the RLB.

The posterior inferior aspect of the right cerebellar hemisphere is supplied by branches of the left PICA. The left PICA also supplies the major part of the inferior aspect of the left cerebellar hemisphere. One of its branches (B PICA) traverses the flocculus, runs dorsal to the facial and vestibulocochlear nerves and then back to the pontomedullary fissure. The AICA on the left side is small and does not reach further than the vestibulocochlear nerve. The right

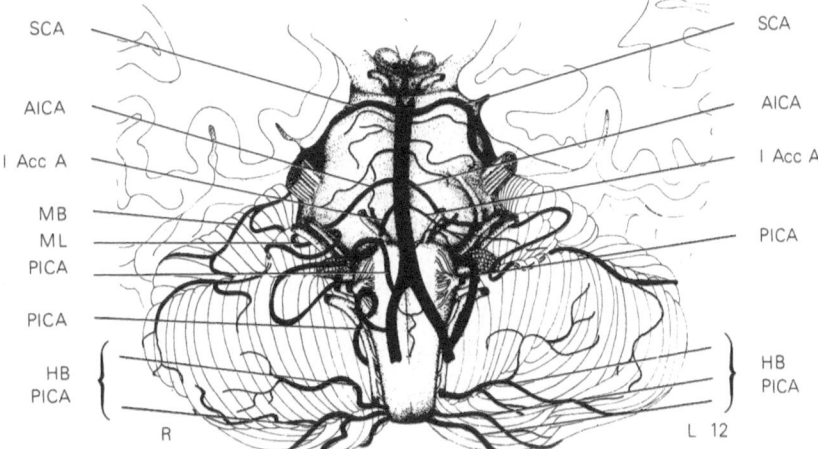

Fig. 13. Case 12

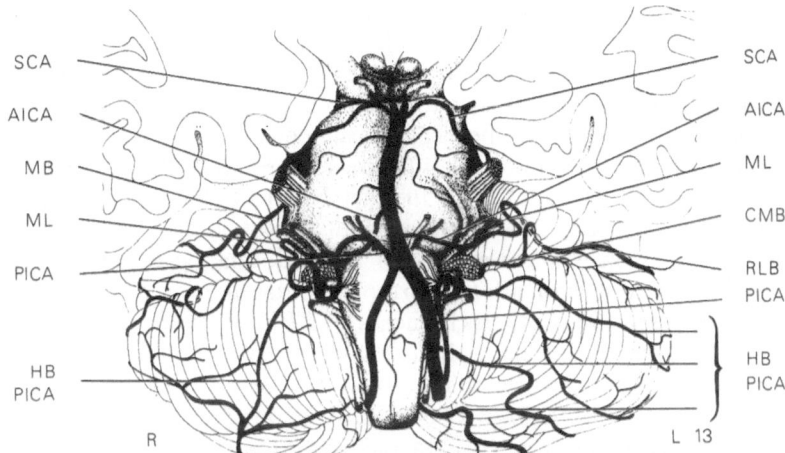

Fig. 14. Case 13

Fig. 15. Case 14

22

SCA bifurcates immediately after originating from the BA. The left SCA is triplicated.

In **case 12** (Fig. 13), the VAs unite caudal to the pontomedullary fissure. On the right side, two PICAs arise from the VA. The posterior inferior aspect of the cerebellar hemisphere is supplied by branches of the PICAs. The right AICA arises from the distal half of the BA. Because of its distal origin, the artery bears no relation to the abducens nerve. The meatal loop (ML) lies dorsal to the facial and vestibulocochlear nerves, but the artery crosses these nerves ventrally. It runs to the horizontal fissure, and its branches anastomose with branches of the PICA and of the marginal branch (MB) of the SCA. Caudal to the origin of the AICA, an inferior accessory artery (IAccA) arises, which crosses the abducens nerve ventrally.

On the left side, branches of the PICA (HB PICA) supply the posterior inferior aspect of the cerebellar hemisphere. The left AICA crosses the abducens nerve and the complex of the facial and vestibulocochlear nerves dorsally and then forms a wide loop that is not related to the internal auditory meatus. The artery passes backward on the surface of the flocculus, and, in the vicinity of the choroid plexus of the fourth ventricle, it changes its direction again and runs laterally in the posterolateral and the horizontal fissures. A small inferior accessory artery (IAccA) crosses the abducens nerve ventrally and ends on the flocculus. The SCAs are symmetrically developed.

In **case 13** (Fig. 14), the VAs join each other near the pontomedullary fissure. On the right side, the PICA arises from the distal part of the right VA near the site of junction with the left VA. In its first trajectory, the PICA runs in a rostrolateral direction. On the surface of the abducens nerve, the artery turns in a caudal direction and after a tortuous course it proceeds between the medulla and the cerebellar tonsil. Its branches (HB PICA) supply the posterior inferior part of the cerebellar hemisphere. The right AICA follows a caudolateral course in its first trajectory. It runs between the abducens nerve and the proximal part of the PICA and follows the pontomedullary sulcus until it reaches the vestibulocochlear nerve. On the ventral surface of this nerve, the meatal loop (ML) is located. The artery proceeds laterally in the horizontal fissure. Its terminal branches anastomose with branches of the marginal branch (MB).

On the left side, the PICA arises from the proximal part of the VA. It runs in a rostral direction parallel to the VA and returns within the medullocerebellar sulcus. The left AICA arises from the proximal part of the BA. It crosses the abducens, the facial and the vestibulocochlear nerves ventrally. The meatal loop (ML) is located lateral to the nerves. Within the suprafloccular part of the horizontal fissure, the artery bifurcates into a caudomedial branch (CMB) and a rostrolateral branch (RLB). One SCA arises from the BA on both sides.

In **case 14** (Fig. 15), the VAs join each other near the pontomedullary fissure. The branches of the right PICA (HB PICA) supply most of the inferior aspect of the cerebellar hemisphere. The AICA follows a caudolateral course, crosses the abducens nerve ventrally and then curls around the facial and vestibulocochlear nerves. The meatal loop (ML) lies dorsal to these nerves. The artery proceeds in the horizontal fissure. No caudomedial branch is present.

The branches of the left PICA (HB PICA) have a similar area of supply to those of the right side. The left AICA arises from about the same level

Fig. 16. Case 15

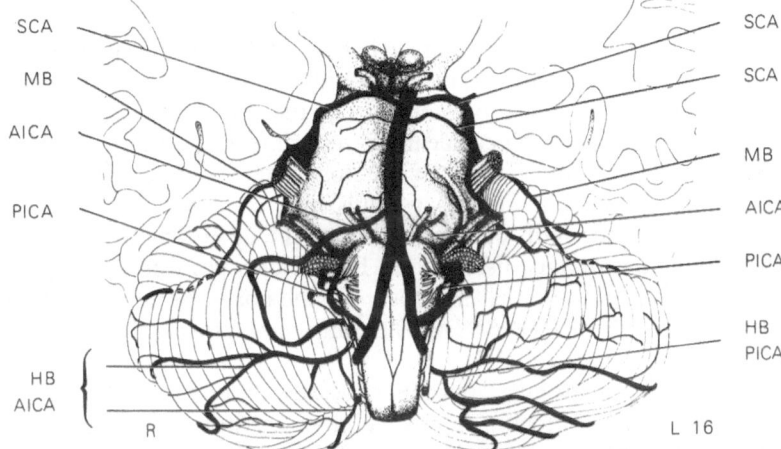

Fig. 17. Case 16

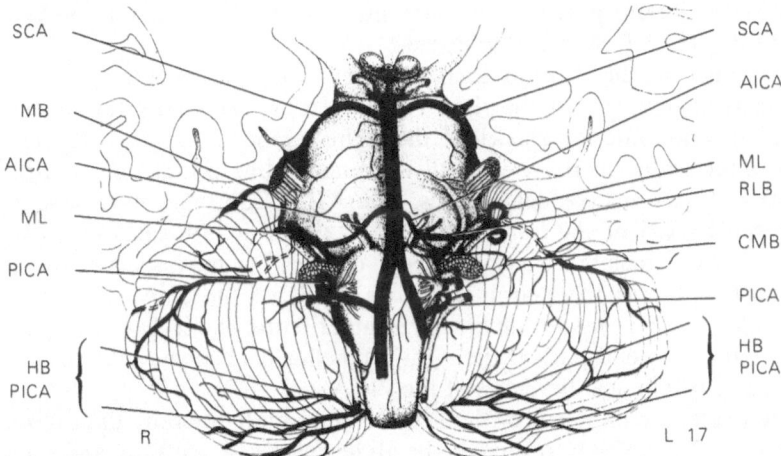

Fig. 18. Case 17

24

of the BA as the right AICA. After having crossed the abducens nerve on the ventral side, the artery forms a loop on the ventral surface of the facial nerve and immediately divides into a caudomedial (CMB) and a rostrolateral branch (RLB). Both branches are only short. The marginal branch (MB) of the SCA runs within the lateral part of the horizontal fissure. One SCA is present on both sides.

In **case 15** (Fig. 16), the VAs unite somewhat distal to the pontomedullary fissure. The branches of the right PICA (HB PICA) supply the posterior inferior part of the cerebellar hemisphere. The right AICA crosses the abducens nerve ventrally and turns toward the internal auditory meatus with a superior concave curve. No real meatal loop is formed. The artery runs laterally in the horizontal fissure.

The branches of the left PICA (HB PICA) supply about the same part of the cerebellar hemisphere as on the right side. The left AICA differs little from the right AICA. It crosses the cranial nerves ventrally. A not very pronounced meatal loop (ML) is formed. The SCAs are approximately symmetrical in their first trajectory.

In **case 16** (Fig. 17), the VAs unite somewhat distal to the pontomedullary fissure. A small PICA is present on the right side. The AICA on this side is much larger. It arises from the proximal third of the BA. The artery crosses the abducens, facial, and vestibulocochlear nerves ventrally, without forming a meatal loop. It courses in the vicinity of the posterolateral fissure, and, just before it reaches the horizontal fissure, it bends in a caudomedial direction. After having formed a wide loop on the surface of the cerebellar hemisphere, it reaches the cerebellomedullary fissure, and here it changes direction again. The artery finally divides into branches that supply the posterior inferior aspect of the cerebellar hemisphere (HB AICA). The marginal branch (MB) of the SCA runs laterally in the horizontal fissure.

On the left side, the anatomic situation is totally different. The PICA is considerably larger, and its branches supply the same territory as the branches of the AICA on the right side. Only a very small left AICA is present, which does not reach further than the complex of the facial and vestibulocochlear nerves. The flocculus is supplied by a branch of the SCA. The marginal branch (MB) runs in the lateral part of the horizontal fissure. One SCA is present on the right side. The left SCA is duplicated.

In **case 17** (Fig. 18), the VAs join each other near the pontomedullary fissure. The hemispherical branches of the right PICA (HB PICA) supply the medial part of the posterior inferior aspect of the right cerebellar hemisphere. The AICA crosses the abducens nerve ventrally. The meatal loop (ML) lies ventral to the facial and vestibulocochlear nerves. The AICA continues its course in the suprafloccular part of the horizontal fissure, where it bifurcates into two branches, neither of which takes a caudomedial course. The most rostral branch reaches almost as far as the marginal branch (MB) of the SCA. The more caudally situated branch runs within the horizontal fissure, and its hemispherical branches supply the lateral part of the right cerebellar hemisphere and anastomose with the hemispherical branches of the PICA.

On the left side, the branches of the PICA extend more laterally than those of the right side. The left AICA crosses the abducens nerve ventrally and then passes between the facial and vestibulocochlear nerves. The meatal loop (ML)

Fig. 19. Case 18

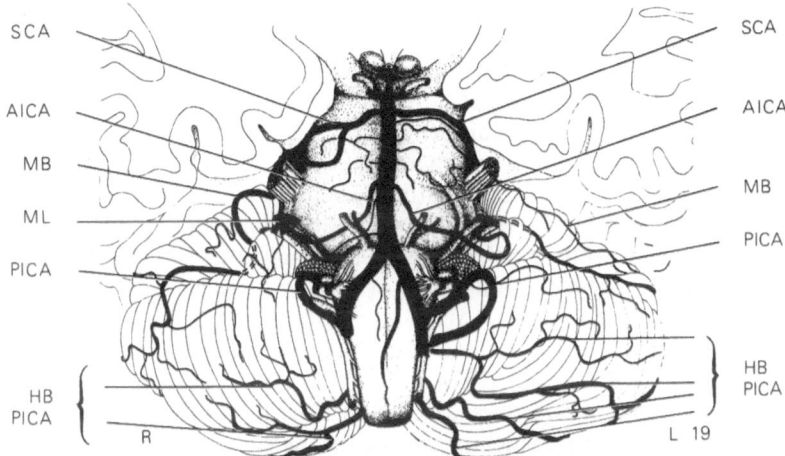

Fig. 20. Case 19

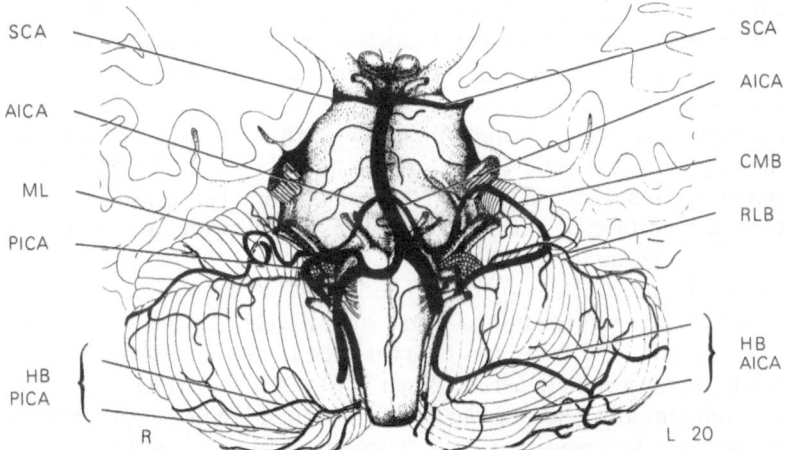

Fig. 21. Case 20

26

is situated immediately dorsal to the nerves. A small caudomedial branch (CMB) is present. One SCA is seen on both sides.

In case 18 (Fig. 19), the arteries show a tortuous course. The VAs unite distal to the pontomedullary fissure. The first part of the right PICA is also tortuous. The hemispherical branches of the right PICA (HB PICA) supply the posterior inferior aspect of the cerebellar hemisphere. The right AICA is small and bifurcates on the ventral aspect of the abducens nerve. The most rostral of the two branches joins the vestibulocochlear nerve. The other branch curves over the flocculus and ends near the posterolateral fissure. The marginal branch (MB) of the right SCA is strongly developed. The branches of the left PICA supply not only the posterior inferior aspect of the cerebellar hemisphere but also the more anterior aspect of the cerebellar hemisphere. One branch of the PICA runs in the posterolateral and the horizontal fissure.

In the suprafloccular part of the horizontal fissure, an artery runs that is an anastomosis between the PICA and the SCA. The left AICA is very small; it crosses the abducens nerve dorsally and does not reach as far as the facial and vestibulocochlear nerves. The left SCA is duplicated and the right SCA is single.

In case 19 (Fig. 20), the VAs unite at the level of the pontomedullary fissure. The branches of the right PICA supply only the medial part of the posterior inferior aspect of the cerebellar hemisphere. The AICA describes a superior concave curve and crosses the abducens nerve dorsally. The meatal loop (ML) lies in close contact with the facial and vestibulocochlear nerves, and the AICA does not proceed more laterally. The right marginal branch (MB) has a tortuous course and is of considerable size. Its branches supply the lateral aspect of the inferior part of the right cerebellar hemisphere and anastomose with the branches of the PICA.

The branches of the left PICA (HB PICA) supply the posterior inferior part of the left hemisphere. The left AICA crosses the abducens nerve ventrally. It runs further in a caudolateral direction, crosses the facial and vestibuloco-chlear nerves also on the ventral side, turns back on the surface of the flocculus, and passes between the nerves to the suprafloccular part of the horizontal fissure. It continues its course in a lateral direction. The marginal branch (MB) is of small caliber. The right SCA bifurcates soon after its origin from the BA.

In case 20 (Fig. 21), the VAs join each other at the level of the pontomedul-lary fissure. The right VA shows a tortuous course. The right PICA arises from the distal part of the VA. It runs caudally within the cerebellomedullary fissure. Its branches (HB PICA) supply the posterior inferior aspect of the right cerebellar hemisphere. The right AICA crosses the abducens, the facial, and the vestibulocochlear nerves ventrally. The meatal loop (ML) lies lateral to the nerves on the surface of the flocculus. The AICA proceeds laterally in the vicinity of the horizontal fissure.

On the left side, no PICA is present. The left AICA arises from the same level of the BA as the right AICA. It first runs parallel to the BA and the distal part of the left VA. Near the choroid plexus of the fourth ventricle, it turns toward the trigeminal nerve. When it reaches the trigeminal nerve, it proceeds caudolaterally and surrounds the flocculus. When it reaches the choroid plexus from the opposite side, it bends in a caudal direction and its branches supply the posterior inferior aspect of the left cerebellar hemisphere.

Just before the artery reaches the choroid plexus, a thin rostrolateral branch arises, which proceeds laterally near the horizontal fissure. One SCA arises from the BA on both sides.

2.2.3 Discussion

In all 20 specimens examined, AICAs were always present. However, in some instances it remains questionable whether a small artery arising from the BA should be regarded as an AICA or as one of the circumferential arteries. This is especially so in cases 16L and 18L (Figs. 17, 19). The AICA always arises from the lower or middle third of the BA, which is in accordance with the findings of others (Stopford 1916a, b; Krayenbühl and Yasargil 1957; Lazorthes et al. 1976; Naidich et al. 1976a, b).

Duplication of the artery was observed in ten cases (Table 1), with a superior accessory artery present in five instances and an inferior accessory artery present in five instances.

Triplication of the AICA was not observed. The relation with the abducens nerve was studied in all cases. In six instances, the artery passes dorsal to the nerve, in 33 instances, the artery passes ventral to the nerve, and in one case (12R: Fig. 13), there was no relation between the abducens nerve and the artery because the position of the AICA was too far rostral. These figures are similar to those of Cushing (1910).

The relation between the AICA and the facial and vestibulocochlear nerves could not be studied because examination took place after removal of the brains from the skulls. For the same reason, no study could be made of the small arteries which arise from the AICA in the vicinity of the internal auditory meatus.

In our material, many variations in the course and areas of supply were observed. Four main types of AICA are distinguished. Fig. 22 is a schematic drawing showing the configuration of these four types. The type I AICA does not reach further laterally than the cerebellopontine angle. The type II AICA extends laterally as far as the lateral border of the cerebellar hemisphere but gives off no, or only a small, caudomedial branch. In the type III AICA, the caudomedial branch is of considerable size and takes over the area of supply of the PICA or part of this area. In this type, the artery also extends laterally, and the lateral segment of the artery is now referred to as the rostrolateral branch. In the type IV AICA, the artery does not extend laterally, but first runs caudolaterally and then changes course in a caudomedial direction. In this case, the marginal branch of the SCA runs in the lateral part of the horizontal fissure. In types III A and IV A, no PICA is present, and the whole inferior aspect of the cerebellar hemisphere is supplied by branches of the AICA.

In Table 1, the most significant data from the 20 specimens examined are summarized. The AICAs are classified according to the four types outlined above. As regards the major branches of the AICA, the nomenclature of Naidich et al. (1976a, b) has, for the greater part, been adopted. The two major branches of the AICA are referred to as the rostrolateral and the caudomedial branches. However, Naidich et al. claim that a caudomedial branch is consistently present and, further, that bifurcation of the AICA always takes place in the vicinity

Table 1. Classification of the anterior inferior cerebellar arteries in the 20 normal specimens

Specimen number	AICA type	Relation to N. Abd.	Meatal loop present	Accessory AICA
1 R	II+CMB	Ventral	+	Inf
L	I	Ventral		
2 R	II	Ventral	+	Sup
L	II	Dorsal		Inf
3 R	IIIA	Ventral	.	
L	II	Ventral		
4 R	II	Ventral	+	Sup
L	II	Dorsal	+	
5 R	I	Ventral		
L	II+CMB	Ventral	+	Sup
6 R	II+CMB	Ventral	+	
L	III	Ventral		
7 R	IVA	Ventral		
L	IVA	Dorsal		
8 R	I	Ventral		Inf
L	II	Ventral		Sup
9 R	II	Ventral		
L	II+CMB	Ventral	+	
10 R	I	Ventral		Sup
L	I	Ventral		
11 R	III	Ventral	+	
L	I	Ventral		
12 R	II	No relation	+	Inf
L	II	Dorsal		Inf
13 R	II	Ventral	+	
L	II+CMB	Ventral	+	
14 R	II	Ventral	+	
L	I	Ventral		
15 R	II	Ventral		
L	II	Ventral	+	
16 R	IV	Ventral		
L	II	Ventral		
17 R	II	Ventral	+	
L	II+CMB	Ventral	+	
18 R	I	Ventral		
L	I	Dorsal		
19 R	II	Dorsal	+	
L	I	Ventral	+	
20 R	II	Ventral	+	
L	IIIA	Ventral		

N. Abd., abducens nerve; Inf, inferior; Sup, superior

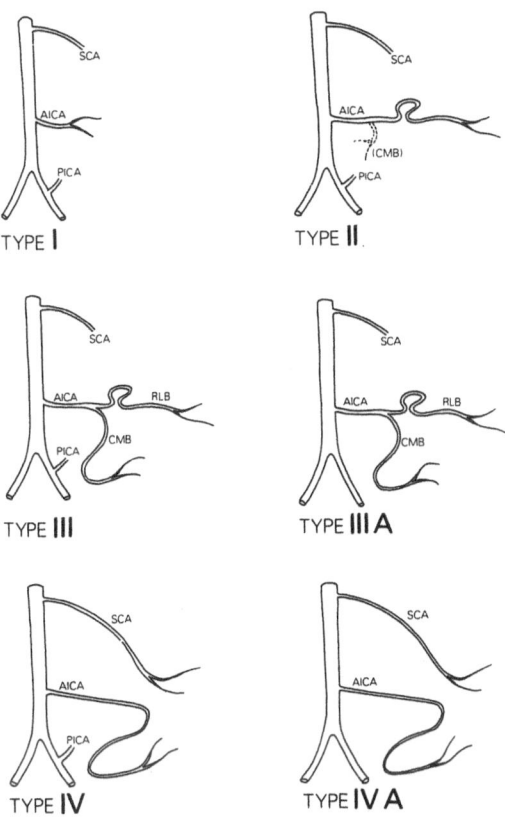

Fig. 22. The main types of the anterior inferior cerebellar artery

of the abducens nerve. This is in contrast with the findings of other authors (Atkinson 1949; Gerald et al. 1973; Perneczky and Tschabitscher 1976), who also describe two main branches but state that bifurcation takes place in the cerebellopontine angle cistern. Naidich et al. also describe this more distal bifurcation but state that it is the division of the rostrolateral artery into the ascending and the descending branch, whereas the caudomedial artery is given off from the AICA more proximally. This finding implies the presence of two major bifurcations and, therefore, of three major branches. A similar differentiation was made by Salamon and Huang (1976), who mentioned three major hemispherical branches of the AICA, but only in cases where no PICA was present.

In the present material, a caudomedial branch as described by Naidich et al. was observed in five instances (3R, 6L, 9L, 11R, and 17L: Figs. 4, 7, 10, 12, 18). In all these cases, the caudomedial branch was given off at, or a few millimeters distal to, the crossing with the abducens nerve. Two of these AICAs are classified as type II (9L and 17L: Figs. 10, 18), because in these instances the caudomedial branch is only a small artery. The other three cases are classified as type III because the CMB has taken over the whole area of supply of the PICA or a considerable part of it.

In one instance (1R: Fig. 2) the CMB emerged from the AICA at its crossing with the facial and vestibulocochlear nerves. In two more cases (5L and 13L:

30

Figs. 6, 14) a branch that followed a caudomedial course emerged even more distally, i.e., from the segment of the AICA that runs in the suprafloccular part of the horizontal fissure. These three cases are all classified as type II because the CMB does not take over a substantial part of the area of supply of the PICA. In the three cases in which a type IV AICA is present (7R, 7L, 16R: Figs. 8, 17), the term caudomedial branch is not used, although the distribution of the artery is comparable with the distribution of the caudomedial branch of the type III AICA. For the same reason, the term rostrolateral branch is not used when no caudomedial branch is present, as is found in the type II configuration of the AICA.

A meatal loop is present in a total of 18 cases. In 16 of these cases, the AICA is of the type II configuration and in one case of the type III configuration (11R: Fig. 12). In one more case (19R: Fig. 20), a meatal loop is present in an AICA of the type I configuration.

The classification chosen by us differs in some respects from the classification of Lazorthes et al. (1950a, b; Lazorthes 1961). The latter is primarily based on the area of supply of the AICA and not on the configuration of the artery. The type III AICA of Lazorthes supplies not only the anterior aspect of the cerebellum but also the anterolateral part of the inferior aspect of the cerebellar hemisphere. Specimens 6L, 12R, 17R, and 19R (Figs. 7, 13, 18, 20) may be considered examples of this type. Our type III AICA is characterized by the presence of two main branches, one of which takes over the area of supply of the PICA or part of it. This type III configuration can be considered as equivalent to the type IV AICA of Lazorthes, because in this case the artery supplies the total anterior and total inferior aspects of the cerebellar hemisphere. Our type IV AICA, which is characterized by its caudomedial extension and by the absence of a lateral branch, and therefore by the absence of a lateral territory of supply, is not mentioned by Lazorthes.

3 RADIOLOGY

3.1 Review of the Literature

3.1.1 Visualization of the Vertebrobasilar System

Visualization of the vertebrobasilar system in a patient was first performed by Moniz and co-workers (Moniz and Alves 1933; Moniz et al. 1933). Contrast medium was injected into the subclavian artery, which was exposed in the supraclavicular fossa. Percutaneous puncture of the subclavian artery was described by Shimidzu in 1937. Lindgren (1950), however, reported poor results with this technique. Direct application of the contrast medium into the vertebral artery was described by Olivecrona (1935), Berczeller and Kugler (1937), and Sjöquist (1938). In these cases, it was necessary to expose the vertebral artery operatively. Percutaneous puncture of the vertebral artery was proposed for the first time by Takahashi (1940), who punctured the artery in its proximal part. Sugar et al. (1949) and Lindgren (1950) punctured the artery in its more distal part, where it courses through the foramina of the transverse processes of the cervical vertebrae. Direct puncture of the vertebral artery was still used by Yasargil in 1962.

Radner (1947) introduced vertebral angiography by means of catheterization. A catheter was inserted into the radial artery and was guided through the main arterial trunk into the vertebral artery. He later described the results of 221 cases performed by this method (Radner 1951). Olsson (1953a, b) catheterized the vertebral artery through the subclavian artery. As in Radner's cases, the artery had first to be exposed operatively.

Percutaneous transbrachial angiography of the vertebral artery was described by Pygott and Hutton (1959), Tatelman and Sheehan (1962), and van den Bergh and van der Drift (1963). Similar catheter techniques were used by Hanafee (1963) and by Roy (1965), who punctured the axillary artery. The subclavian artery was punctured by Amplatz and Harner (1962) and by Weibel (1966).

In 1953, Seldinger introduced a new catheter technique. Via an arterial puncture, a guide wire was placed into the vessel. Over this guide wire a catheter was introduced. Using this technique, it was possible to introduce a catheter with a wider diameter than the needle used for the puncture. Catheterization of the vertebral artery through puncture of the femoral artery was done by Lindgren (1956). However, he did not use the method described by Seldinger but punctured the femoral artery with a thick needle, through which a thin catheter was introduced. Bonte et al. (1958) used the Seldinger method to catheterize the vertebral artery. Soon other reports using the same method were published (Brenner 1961; Cronquist 1961; Scatliff et al. 1965; Newton et al. 1966; Braun et al. 1966). Scatliff et al. (1965) formulated some disadvantages of the catheter method. They regarded plugging of the vertebral artery by the catheter as the principal danger and, because of this, they emphasized that the catheter approach is not the perfect answer for angiography of the vertebro-basilar system.

Angiography of the vertebrobasilar system is nowadays routinely done using the Seldinger method and the femoral approach. Thin catheters with an outer size of 5.0 F or even 4.1 F, are used to catheterize selectively the largest of the two vertebral arteries. Fluoroscopy is used in a test injection to ensure correct positioning of the catheter and free flow of the contrast medium. Guide wires should be handled with great care: When they are removed, no injection should take place before blood drips from the catheter, otherwise air may be sucked into the catheter.

Attention should be paid to the reactions of the patient: If there are complaints which could even remotely be attributed to the position of the catheter or reaction to the injection of contrast material, the examination should not be carried out if no vital indication exists. In the latter case, it may be possible to introduce the catheter into the vertebral artery on the contralateral side. Directly after the injection of contrast medium, the catheter should gently be pulled out of the vessel to prevent spasms.

The contrast medium that is used must be of low toxicity, as most modern contrast media are. In critical cases, a nonionic, water-soluble contrast medium should be used. If all these conditions are fulfilled, vertebral angiography may be considered a low-risk investigation.

3.1.2 Subtraction

Visualization of the smaller intracranial vessels is only satisfactory when subtraction is used. This is especially the case with respect to the smaller branches of the BA. Tönnis and Schiefer (1959) stated that the small arteries of the pons, the IAA and the AICA, have no angiographic significance, because in the lateral projection these arteries cannot be seen through the dense bone structures. Indeed, the AICA and its branches are completely obscured by the petrous bones in the lateral projection if no subtraction is used. It may also be difficult in the anteroposterior (AP) projection to identify the exact course of the AICA and its branches if subtraction films are not used.

The principle of subtraction was first used in astronomy (Pickering 1904). This was noted by Ziedses des Plantes (1934), who introduced the use of subtraction to medicine. However, it was many years before the technique was routinely used in angiography. In 1961, in a detailed monograph by Ziedses des Plantes, some fine examples of subtraction in angiography were shown. In this monograph, the principle of so-called second-order subtraction was also discussed.

The use of this technique in carotid angiography was demonstrated by Hanafee and Shinno (1966). They concluded that second-order subtraction permits visualization of the brain substance in the capillary phase. Avascular areas, localized cerebral edema, and minor vascular blushes can be better demonstrated by this method than by first-order subtraction. In a paper on the radiographic anatomy of the AICA, Takahashi et al. (1968) reviewed 250 vertebral angiograms. They frequently used second-order subtraction, which was considered essential for the adequate visualization of the finer ramifications of the AICA.

In order to obtain a first-order subtraction, an "empty" roentgenogram without contrast is required. This roentgenogram is usually obtained immediately prior to serial angiography. A diapositive is made of this "empty" film and the diapositive is then superimposed on one or more roentgenograms of the angiographic series. On the contact print of these two roentgenograms, the bony structures are "grayed out" and the arteries which are filled with contrast are enhanced. On this film only a faint outline of the bony structures remains. If this faint image of the skull still interferes with the angiography, a second-order subtraction can be made. A second diapositive is obtained by registering the first "empty" roentgenogram with its own diapositive. The faint film thus produced is, together with the first diapositive, superimposed on the roentgenogram of the angiographic series. Contact prints of these three films are called second-order subtractions and show almost no bony structures.

3.1.3 The AICA in Vertebral Angiography

Many reports on vertebral angiography were published during the 1950s and 1960s. No doubt existed as to the significance of this technique in the visualization of aneurysms (Collins et al. 1957; Krayenbühl and Yasargil 1957). The importance of vertebral angiography in the diagnosis of posterior fossa tumors has been emphasized by many authors (Columella and Papo 1955; Löfgren 1956; Isfort 1960; Brenner 1961; Mones 1961; Allcock 1962; Yasargil 1962). Avascular areas and abnormal tumor stains were regarded as useful signs, but displacement of arteries was also considered to be of value in localizing tumors. Olsson (1953) regarded such tumor diagnosis by displaced vessels as disappointing. He stated that the vertebral system is a sector with a much more primitive pattern than the carotid system, and, unlike the carotid system, does not bear a regular relation to the midline. Moreover, the vascular pattern varies widely from side to side and from patient to patient. Olsson was of the opinion, however, that experience and a more detailed knowledge of the anatomy would in time make it possible to judge displacements of various types. Khilnani and Silverstein (1963) shared Olsson's view that vascular displacements in the posterior fossa are only rarely of localizing value.

Those authors who regarded displacement of posterior fossa arteries as an important localizing sign only mentioned the BA, the PICA, and the SCA. The AICA was considered by them to be of no angiographic significance (Radner 1951; Tönnis and Schiefer 1959). Allcock (1962), who regarded vertebral angiography as a useful tool in the diagnosis of space-occupying lesions, did not mention the AICA in a review of 96 cases: Radiological visualization of the AICA was described for the first time in the late 1960s (Dilenge and David 1967; Mani et al. 1968; Takahashi et al. 1968; Ziedses des Plantes 1968; Peeters 1969). Apparently, this was due to a more widespread use of subtraction techniques. Although knowledge of the anatomy of the AICA was available at the beginning of the century, it is surprising that, in most angiographic reports, little attention is paid to the variable origin, course, and distribution of this artery.

The following data on the radiographic anatomy of the AICA have been mentioned in the radiological literature. Salamon and Huang (1976) described the loop that is formed by the AICA where it is in close relation to the flocculus. The variable size of the artery and the caudal extension of the AICA when the PICA is atretic or absent were also reported.

Takahashi et al. (1968, 1974) analyzed 100 angiographies and found that the AICA followed the "usual course" in about half of the cases. In 40%, one or both AICAs supply the areas which are normally supplied by the PICA. The reverse situation, in which the PICA supplies the area normally supplied by the AICA, was only present in 9% of the cases.

Gerald et al. (1973) found the AICA was equal to or larger than the PICA in 25% of the cases. In 60 of 74 AICAs visualized, a meatal loop was seen. It was not possible to determine whether this loop was inside or outside the internal auditory meatus. Gerald also described two main branches of the AICA and observed these two main branches in 72 cases. In 18 cases only, branching was not seen, with the medial branch usually absent. The course of the proximal segment of the AICA in angiography may be straight, convex superiorly, convex inferiorly, or the artery may even show a double loop with both superior and inferior convexities.

Ross and Du Boulay (1976) pointed out that usually the proximal or pontine portion of the AICA shows an inferior convexity, but that this portion is often tortuous when the BA curves to the ipsilateral side.

An inverse relation in size between the AICA and the PICA is mentioned by George (1974). In the same report, a common trunk of the AICA and the PICA is mentioned, which is present in about 30% of the cases. George also describes the meatal loop and the medial branch, which is infrequently visualized. He distinguishes four segments of the AICA: a pontine segment, a cerebellopontine angle segment, a flocullar segment, and a semilunar segment. The AICA is considered by him the diagnostically most important vessel for the evaluation of cerebellopontine angle tumors, and he emphasizes that preoperative visualization of the artery is important in order to avoid inadvertent ligation of the AICA during acoustic neurinoma surgery.

In 1949, Atkinson pointed out the danger of occlusion of the AICA during surgery. This was stressed again by Takahashi et al. (1968), and by Pinto et al. (1977), in a paper on the base view in vertebral angiography, in which the importance of preoperative visualization of the cerebellar vasculature and especially of the AICA was emphasized. Kieffer et al. (1975), however, delineated the AICA in only 10 of 19 cases of cerebellopontine angle tumors, and they regarded the artery as usually small and difficult to recognize without subtraction and magnification techniques.

The most important study of the radiological appearance of the AICA in the lateral projection is that of Naidich et al. (1976a, b). The angiographic anatomy of both main branches of the AICA (the rostrolateral artery and the caudomedial artery) is described in detail and illustrated by drawings and photographs of anatomic specimens. Both main branches are divided into segments.

The loop that is formed by the rostrolateral artery in the vicinity of the porus acusticus internus is either single or double. The single loop is designated the meatal loop, and the double loop is called the M-segment because of its resemblance to the letter M in the lateral projection. This M-segment consists of a meatal loop and a brachial loop, the latter being in close relation to the brachium pontis. Together with the suprafloccular segment of the rostrolateral artery, the brachial loop of the M-segment forms the brachial segment of the rostrolateral artery. The rostrolateral artery divides into an ascending and a descending branch. A portion of the descending branch is designated the retrofloccular segment, and this segment, together with the suprafloccular segment and the M-segment, forms the so-called floccular loop. The distal part of the descending branch is called the biventral segment. A part of the caudomedial artery is also called the biventral segment by Naidich et al. This segment represents the portion of the caudomedial artery located distal to a lateral loop formed by the more proximal part of the caudomedial artery.

Angiographic demonstration of the internal auditory artery is seldom mentioned as the IAA is in most cases a branch of the AICA. Walker (1965) was able to visualize the internal auditory artery in dogs. However, he doubted whether it would be possible to do so in man. According to Sartor (1976) and Ross and Du Boulay (1976), the internal auditory artery is only occasionally visualized in vertebral angiography. Smaltino et al. (1971) found that the internal auditory artery in vertebral angiography is visible only for a short interval of time. They succeeded in visualizing the internal auditory artery in 22 of 30 cases. In one-third of the cases, the artery was observed bilaterally. They subdivided the internal auditory artery into a precanalicular, or cisternal, segment and an intracanalicular segment. The artery was usually visible for a length of about 2 cm and had a diameter of 0.5 mm. This artery was also studied in cases of "sudden deafness" and Ménière syndrome in which only the stump of the so-called cerebellolabyrinthine artery was visible. In cases of intra- and extracanalicular neurinoma of the eight nerve, the intracanalicular segment was stretched and the cisternal segment stretched and displaced.

Vogelsang (1974) also noted that the internal auditory artery is only visible during a short period and mostly at the beginning of the arterial phase. Rapid film exchange is therefore necessary. Subtraction and magnification techniques are also considered essential by him. When these conditions are met, it is possible to visualize the internal auditory artery on one side in about half of the cases and bilaterally in one-third of the cases.

3.1.4 Projection

Most authors consider the transfacial or straight AP view the best projection to visualize all segments of the AICA in the frontal view (Dilenge and David 1967; Greitz and Lindgren 1971; Takahashi et al. 1968, 1974; Hanafee and Wilson 1972; Sartor 1976; Krayenbühl et al. 1979). Others routinely use Towne or half-axial views (Goree et al. 1964; Gerald et al. 1973; Naidich et al. 1976a, b; Taveras and Wood 1976). Occasionally, it may be difficult to identify the exact course of the AICA in these projections because of the superimposed branches of the PICA (Takahashi et al. 1968, 1974; Krayenbühl et al. 1979).

36

A disadvantage of the AP projection is that the caudomedial branch is greatly foreshortened (Ross and Du Boulay 1976). Furthermore, posterior displacement of the AICA may be better visualized using the Towne projection (Krayenbühl et al. 1979).

Little attention has been paid to the AICA in the lateral projection. Even after subtraction, it may often be impossible to identify the artery through the dense structures of the petrous bones (Sartor 1976; Takahashi et al. 1968). Also, superimposition of the right and left AICAs often makes exact identification of the artery difficult. The extensive study of the AICA in the lateral projection by Naidich et al. (1976a, b) has already been mentioned. The anatomic description of both main branches of the AICA in this study is very detailed and many of the details can be recognized in the lateral projection. No figures are presented on the percentage of angiograms in which this detailed analysis was not possible because of superimposition of either bony structures or of the contralateral AICA.

The oblique projection in the diagnosis of cerebellopontine angle tumors has been discussed by Hayman et al. (1979) and was mentioned by Huang et al. (1974) and Moscow and Newton (1975). In order to obtain this projection, the face of the patient is turned 45° toward the side of interest, and the central ray is directed through the long axis of the cerebellopontine angle cistern. The vessels on the surface of the pons, the medulla, and the cerebellar hemisphere on this side are seen in profile. Superimposition of bone and vessel images on the tumor site is thus reduced to a minimum. Furthermore, problems caused by variation in vascular anatomy are also reduced. The authors state that variations in origin and branching pattern of vessels such as the AICA are of secondary importance. However, the oblique projection can provide important information about tumor vascularity and can make vital brain stem vessels visible, which will reduce the risk of operation.

The base view is not often used although it offers certain advantages (Malter and Roberson 1972; Pinto et al. 1977). In particular, medial displacement of the so-called prepontine segment of the AICA by a cerebellopontine angle tumor, indicating brain stem impingement, can be made visible in the base projection. This observation may imply adherence of the tumor to the anterolateral aspect of the brain stem, making total excision of the tumor impossible. Other projections do not provide this important information.

3.1.5 Vertebral Angiography in Posterior Fossa Tumors: General Aspects

The radiological diagnostic approach to posterior fossa tumors has been entirely changed by the introduction of CT scanners. With the present third or fourth generation CT scanners, small acoustic neurinomas can be diagnosed, even when these tumors are located within the internal auditory canal (Valavanis et al. 1982). The rapid development of CT technology has drastically diminished the importance of angiography in locating both supratentorial and infratentorial tumors (Pertuiset et al. 1979; Philippon et al. 1979).

Diagnosis of space-occupying lesions in the posterior cranial fossa by means of angiography alone has always been considered difficult (Greitz and Lindgren 1971), and some authors have even doubted the diagnostic value of vascular displacements in vertebral angiography (Olsson 1953; Khilnani and Silverstein 1963). It may nevertheless be stated that the numerous reports on vertebral angiography in posterior fossa tumors which appeared during the years that CT scanning was not possible have provided valuable data on displacements of posterior fossa arteries in various space-occupying lesions and on the vascularization of tumors. Nowadays, the localization of posterior fossa tumors by means of CT is superior to localization by means of vertebral angiography.

However, angiographic findings may still be important, if not for the localization of tumors, for the visualization of tumor vasculature and the relation of the tumor to the posterior fossa vessels.

There are several localizing classifications of posterior fossa tumors. Naidich et al. (1976) distinguished between extra-axial and intra-axial tumors. Extra-axial tumors may be subdivided into: (a) clival tumors, (b) cerebellopontine angle tumors, (c) tentorial tumors, and (d) infratentorial meningeomas. Intra-axial tumors comprise: (a) brain stem gliomas, (b) cerebellar peduncular and superior hemispherical tumors, and (c) inferior hemispherical and intra-fourth-ventricular tumors. Taveras and Wood (1976) also made a division into extra-axial and intra-axial tumors. Furthermore, they divided the posterior fossa into an anterior and a posterior compartment. The same was done by Greitz and Lindgren (1971). The posterior compartment contains the cerebellar vermis and the cerebellar hemispheres, whereas the anterior compartment contains the brain stem, the cerebellopontine angles, and the clivus.

Peeters (1969) subdivided the posterior fossa tumors into: cerebellopontine angle tumors; brain stem tumors; tumors in the region of the aqueduct, fourth ventricle, and cerebellar vermis; tumors of the cerebellar hemisphere; and clivus tumors. A similar subdivision was made by Takahashi (1974) and also by George (1974), who, in addition, distinguished a category of rare extra-axial tumors, such as trigeminal neurinomas, glomus jugulare tumors, and lesions of the foramen magnum and of the floor of the posterior fossa.

It may be pointed out that not all these tumors will effect the course of the AICA in the same way or to the same extent.

3.1.6 The AICA in Cerebellopontine Angle Tumors

Masses arising in the cerebellopontine angle are most frequently acoustic neurinomas. Other space-occupying lesions causing cerebellopontine angle symptoms may be meningeomas, epidermoid tumors, arachnoid cysts, aneurysms, tuberculomas, gliomas, or glomus jugulare tumors (Wende and Nakayama 1972). High-resolution axial and coronal CT and oxygen CT cisternography are currently the methods of choice in the diagnosis of acoustic neurinomas (Valavanis et al. 1982). However, there are several reasons to perform vertebral angiography in the preoperative evaluation of cerebellopontine angle tumors. Preoperative angiography is considered essential in order to exclude the presence of vascular lesions like aneurysms causing cerebellopontine angle signs (Castaigne et al. 1967; Long et al. 1973; Bender 1973; Taveras and Wood 1976; Johnson and Kline 1978). Such aneurysms of the AICA may even be located within the internal auditory canal (Hori et al. 1971). Furthermore, tortuous arteries of the vertebrobasilar system may cause cranial nerve signs simulating a cerebellopontine angle tumor (Kerber et al. 1972; Shalit and Reichenthal 1978). Philip-pon et al. (1979) stated that in cerebellopontine angle tumors, CT provides enough information for surgery in those cases in which enlargement of the internal auditory meatus can be seen on the skull film. However, when the clinical picture is incomplete or the CT scan atypical, vertebral angiography is indicated.

Another major reason to perform vertebral angiography preoperatively with cerebellopontine angle tumors is to visualize the relation of the major vessels (one of which is the AICA) to the tumor, as well as the vascularity of the tumor itself (Hitselberger and House 1966; Valvassori 1969; Takahashi et al. 1971; George 1974; Hakuba 1977; Pinto et al. 1977; Numaguchi et al. 1980). Knowledge of the relation between the tumor and the major vessels in the vicinity can be of vital importance.

Atkinson (1949) was the first to point out the danger of damage to the AICA during cerebellopontine angle surgery. Seven cases were described in which infarction of the lateral tegmental region of the pons had occurred during, or directly after, the operation. At autopsy, occlusion of the AICA was found in all cases in the neighborhood of the internal auditory meatus, due either to a clip placed on the AICA or to postoperative thrombosis of the artery. This suggests that occlusion of the AICA may be a cause of death following cerebellopontine angle surgery, especially when the AICA has a larger territory of supply than usual, as may be the case when the PICA is small. Atkinson considered a rise in blood pressure an important warning sign.

Hitselberger et al. (1966) reported a series of 114 patients who were operated on because of an acoustic tumor. Vital sign changes, of which elevation of blood pressure was the most common, were exhibited by 34 patients. Pulse changes were variable; irregular respiration and either apnea or hyperpnea were also noted. These changes in vital parameters were regarded as a warning that further attempts at tumor removal might be dangerous. In one patient who died, the upper third of the pons and the middle cerebellar peduncles were found to be necrotic. As no occlusion of arteries was found, it might be assumed that arterial spasm was the cause of infarction. The authors further pointed out that in the case of dominance of the AICA, the surgeon should be especially alerted to severe complications.

Olivecrona (1967) made similar observations and showed that when the tumors are larger, more pontine infarctions occur. One of the three patients described by Perneczky et al. (1981) developed a caudolateral pontine syndrome after acoustic neurinoma surgery. On re-exploring the operating field, the caudo-medial branch of the AICA was found to be spastic. After application of papav-erine, the vessel was released, and the symptoms diminished.

As already noted, vascular displacements by cerebellopontine angle tumors can be of localizing value but have generally been considered less important since the introduction of CT scanners. Takahashi et al. (1971) found that the diagnostic accuracy of vertebral angiography is low for tumors which are smaller than 1.5 cm in the greatest diameter. Earlier reports on vertebral angiography in cerebellopontine angle tumors did not mention displacements of the AICA. The most important sign of a space-occupying lesion in the cerebellopontine angle was upward displacement of the SCA (Olsson 1953; Isfort 1960; Leman et al. 1967a, b, c; Dilenge and David 1967). Displacements of the PICA and the BA were also mentioned in these papers. Smaller arteries, usually branches from the AICA, surrounding a cerebellopontine angle tumor have been mentioned in several reports (Goree et al. 1964; Dilenge and David 1967; Ziedses des Plantes 1968; Greitz and Lindgren 1971; Moscow and Newton 1975).

In a number of more recent reports, displacement of the AICA in angiography is regarded as an important localizing sign. Since the artery lies in close contact with either the ventral or the dorsal aspect of the facial and vestibulo-cochlear nerves, the vessel may be either depressed or elevated by a tumor (Takahashi et al. 1968, 1971; Peeters 1969; Hanafee and Wilson 1972; George 1974; Kieffer et al. 1975; Krayenbühl et al. 1979). Numaguchi et al. (1980) reported 42 patients with acoustic neurinomas and found superior displacement in 25 cases and inferior displacement in 12 cases.

In general, superior displacement of the AICA is regarded as a reliable indication of the presence of a cerebellopontine angle tumor. Since the AICA usually takes a downward course, inferior displacement can only be appreciated in larger tumors (Hanafee and Wilson 1972; Takahashi et al. 1971).

Pinto et al. (1977) pointed out the posterior displacement of the AICA away from the petrous bone. This type of displacement is seen more frequently than a superior or an inferior shift with acoustic neurinomas. The authors considered this type of displacement a key sign in the differentiation between intra- and extra-axial tumors. It can be seen especially in the base view.

Naidich et al. (1976a, b) described an arcuate displacement of the main trunk of the AICA and of the rostrolateral branch in the lateral projection. Because the so-called M-segment is located posteroinferior and medial to the porus acusticus, the usual acoustic neurinoma that grows into the angle from the internal auditory canal displaces the AICA and its rostrolateral branch posteroinferiorly and medially. The part of the artery that is located proximal to the porus and that is anchored at both the BA and the porus becomes bowed and elongated. Inferior displacement of the caudomedial branch may also be seen.

Ziedses des Plantes (1968) mentioned that most cerebellopontine angle tumors do not show a significant staining. Hanafee and Wilson (1972) found that approximately one-third of patients with cerebellopontine angle tumors have some degree of tumor staining. Numaguchi et al. (1980) could demonstrate tumor stains in 38% of patients with acoustic neurinomas. They found it difficult to differentiate between meningeomas and acoustic neurinomas by tumor stains only. Takahashi et al. (1971) found tumor staining in 20% of cerebellopontine angle tumors. These stains, which were supplied by the AICA, were reported to be faint and patchy. Kieffer et al. (1975) also regarded the AICA as usually the chief artery supplying extra-axial and intra-axial tumors in the cerebellopontine angle.

Levine et al. (1973) demonstrated a contribution to the vascular tumor supply from the external carotid artery in two cases of acoustic neurinoma. In one of these cases, the tumor was supplied exclusively by the external carotid artery, and there was no apparent contribution from the vertebrobasilar circulation. The authors suggested the study of external carotid circulation in the radiological evaluation of acoustic neurinomas.

Moscow and Newton (1975) found contributions from neighboring meningeal vessels to tumor vascularization in one-third of cases.

Théron and Lasjaunias (1976) reported six cases of acoustic neurinomas supplied by branches of the external carotid artery. In three of these cases only, injection of the VA partially opacified the tumor. The internal carotid artery contributed to the blood supply of the tumor in three cases via the lateral artery of the clivus. Superselective catheterization of the internal maxillary artery and of the ascending pharyngeal artery was performed, and tumor staining was observed in five and two cases, respectively. Superselective catheterization of the occipital artery in three cases did not reveal tumor staining. The authors suggested a new angiographic protocol for the study of cerebellopontine angle tumors. This protocol should include: (a) selective injection of the vertebral artery, (b) selective injection of the internal carotid artery, and (c) selective injection of the main stem of the external carotid artery. Complementary superselective injections should then be made, if necessary, of the internal maxillary artery or the middle meningeal artery, the ascending pharyngeal artery, and the occipital artery. In the case of a hypervascular tumor, preoperative embolization may be considered.

Perneczky (1980) stated that there is no evidence that acoustic neurinomas are primarily supplied by the AICA. The primary tumor site is outside the arachnoid. The AICA is easily stripped away from the tumor surface at operation, since it only sends a few minute branches into the tumor. When the tumor is split into a medial and a lateral part, its lateral portion appears to bleed much more than its medial portion. Therefore, the blood supply of the tumor is primarily from dural vessels.

3.1.7 The AICA in Cerebellar Tumors

The tumors that most frequently occur in the cerebellum are medulloblastomas, hemangioblastomas, astrocytomas, ependymomas, and metastatic tumors. As with cerebellopontine angle tumors, CT is nowadays the method of choice in the diagnosis of cerebellar tumors. In the years before CT scanning became a routine diagnostic procedure, important localizing value was attributed to displacements of the various segments and branches of the PICA (Wolf et al. 1962; Yasargil 1962; Dilenge and David 1967; Hoffmann and Leifer 1967; Peeters 1969; Greitz and Lindgren 1971; George 1974; Taveras and Wood 1976).

With tumors located within the superior part of the cerebellar hemisphere, the SCA may be stretched and elevated (Wolpert 1971). The course of the AICA is affected by cerebellar masses, but these changes are considered unspecific (George 1974).

Takahashi et al. (1974) mentioned a straightened and stretched course of the AICA when fourth ventricle or cerebellar tumors are associated with increased pressure within the posterior fossa. This appearance may simulate elevation or depression of the artery. Elevation or depression of the AICA may also exist when a cerebellar tumor extends within the cerebellopontine angle.

Naidich et al. (1976a, b) made a more detailed analysis of displacements of the AICA in cerebellar tumors. According to their observations, the main trunk of the AICA and the so-called ascending artery may be displaced by vermian and medially situated quadrangular lobule tumors. Moreover, superior and inferior hemispherical masses can be distinguished, because the rostrolateral branch of the AICA, which courses in the horizontal fissure and thus describes an equator through the cerebellar hemisphere, will be bowed inferiorly or superiorly, respectively, by these masses. The brachial segment of the rostrolateral branch of the AICA will be displaced antero-inferolaterally by lesions extending through the peduncle. The loops that are part of the so-called M-segment may be compressed in the lateral projection by tumors within the superior part of the cerebellar hemisphere. Inferior cerebellar masses may cause stretching and antero-inferior bowing of the biventral segment of the caudomedial branch of the AICA.

Naidich et al. also made the observation that in patients with small or absent PICAs, evaluation of tonsillar herniation can be made from the position of the biventral segment of the caudomedial branch of the AICA. The lateral loop of the caudomedial branch can be a useful indicator of the mass effect of the antero-inferomedial aspect of the cerebellar hemisphere.

3.1.8 The AICA in Brain Stem Tumors

Tumors in the brain stem usually arise in the pons or at the pontomedullary junction and may extend into the midbrain or into the medulla oblongata. Gliomas are the most common brain stem tumors, but metastases and medulloblastomas may also arise in the brain stem. As with other posterior fossa tumors, CT will be the method of choice in the diagnosis of brain stem masses. Doubts about the value of vertebral angiography in the diagnosis of brain stem tumors have been expressed by some authors (Lefèbvre et al. 1963).

Apart from venous displacements, the most important angiographic changes in the presence of a brain stem tumor are the following:

Forward (and occasionally backward) displacement of the BA

Stretching and posterior displacement of the posterior medullary, supratonsillar, and retrotonsillar segments of the PICA

Lateral arching of the perimesencephalic portion of the SCA, when the tumor extends into the midbrain (Peeters 1968, 1969; Huang and Wolf 1970; Wolpert 1971; Greitz and Lindgren 1971; Dilenge and David 1967; Seeger and Gabrielsen 1972; Takahashi 1974)

Abnormalities in the course of the AICA are best seen in the AP view. The proximal section of the artery may be stretched (George 1974) or inferiorly displaced (Peeters 1969; Seeger and Gabrielsen 1972; Takahashi 1974). In the lateral view, there are in most cases no dramatic changes in the course of the AICA (Naidich et al. 1976a, b).

3.1.9 The AICA in Ischemic Lesions

A case of occlusion of the right AICA was reported by Goodhart and Davison in 1936. The patient was hypertensive and showed ataxia in the finger-to-nose test, more marked on the right side. There was a generalized weakness of all muscle groups and all reflexes were slightly hyperactive. A supranuclear weakness of the left side of the face was present. At autopsy, apart from the occlusion of the right AICA, numerous atheromatous plaques were found in the basilar, vertebral, and cerebellar arteries. Several small areas of softening were present in the thalamus, putamen, corpus callosum, and right pallidum. Furthermore, an area of softening was present in the upper third and on the inferior surface of the lobulus paramedianus and lobulus ansiformis crus 2 on the right side. No ischemic lesion of the pons was reported. However, because widespread lesions were present throughout the brain, this case cannot be regarded as a typical example of occlusion of the AICA.

Adams (1943) stated that the clinical picture following occlusion of the AICA is probably the least known of any of the cerebellar artery syndromes. He also described a case of complete occlusion of the right AICA. The right flocculus, the right biventral lobule, and the right superior and inferior semilunar lobules were softened. An area of infarction was also present in the right half of the basal part of the pons. The right middle and inferior cerebellar peduncles were also softened. The clinico-anatomic correlation in this case was obvious: Palsies of the facial and vestibulocochlear nerves existed as a result of the involvement of these nerves and their nuclei. Vertigo, nausea, and vomiting were ascribed to lesions of the vestibular nuclei and their connections with the nuclei of the vagus and oculomotor nerves. Ipsilateral loss of pain and temperature sensation of the face was caused by interruption of the spinal tract and the nucleus of the trigeminal nerve. Absence of contralateral hypalgesia and thermohypesthesia was thought to be due to the extreme lateral and posterior position of the lesion, which spared most of the spinothalamic tract. The ipsilateral Horner's syndrome was related to interruption of pupillodilator fibers, which descend ipsilaterally from the hypothalamus and converge in the lateral portion of the pons and medulla. The ipsilateral signs of cerebellar ataxia resulted from lesions in both the spinocerebellar and pontocerebellar systems.

According to Adams, the onset of the so-called "syndrome of the AICA" is usually sudden and unaccompanied by loss of conciousness. Vertigo, nausea, and vomiting are the first and most important symptoms. Other symptoms appear within a few hours. The clinical course is rarely fatal, and usually the patient gradually improves. Signs pointing to involvement of the corticospinal tracts and medial lemnisci are absent. The symptoms and signs are chiefly related to softening of the lateral portions of the brain stem and cerebellar peduncles rather than to involvement of the cerebellar hemisphere. Adams pointed out that because of the variability of the artery, probably not every case of occlusion of the AICA will be identical to the one described.

Adams and Victor (1977) and Brouckaert et al. (1981) made a distinction between a minimal and a maximal syndrome of infarction in the AICA territory, thus emphasizing the importance of the variability of the artery.

The papers by Atkinson (1949), Hitselberger et al. (1966), Olivecrona (1967), and Perneczky et al. (1981) have already been mentioned with respect to cerebellopontine angle surgery. Infarction of the lateral pontine tegmentum can result from occlusion of the AICA. One of the cases reported by Perneczky et al. is an example of this. The other two patients they describe suffered from a spontaneous occlusion of the AICA. Three anatomic factors are considered of importance:
1. The perforating branches of the AICA, coursing to the lateral pontomedullary region, supply the nuclei of the trigeminal and facial nerves, the medial lemniscus, the ventral tegmental tract, and the spinothalamic tract.

2. The labyrinthine artery is a functional end artery, and lesions of the AICA will therefore be associated with partial or complete functional loss of the vestibulocochlear system.
3. There are many anastomoses between the hemispherical branches of the AICA and the other cerebellar arteries, and the cerebellar hemisphere will therefore not be easily infarcted.

The arterial supply of the pons has been studied extensively by many investigators. In accordance with Foix et al. (1925), Hiller (1952) distinguished three major arterial zones, supplied by the paramedian, the short circumferential, and the long circumferential arteries. Three major pontine ischemic syndromes can thus be distinguished: the paramedian, the lateral, and the tegmental pontine syndromes. The first is caused by occlusion of one or more paramedian branches and is characterized by crossed hemiparalysis, rarely accompanied by hemianesthesia. The patient looks away from the focus. The lateral pontine syndrome results from occlusion of one or more of the short circumferential branches. It is rare in its pure form and consists of an ipsilateral cerebellar disturbance, sometimes combined with a transient dissociated sensory deficit.

The nutritive arteries of the caudal portion of the pontine tegmentum originate as smaller branches of the short circumferential pons arteries and of the AICA. The symptoms of lasting ischemia of the AICA territory are variable and may include ipsilateral dyssynergia, ataxia, hypoesthesia in the face, peripheral facial paralysis, deafness, and a positive Horner's sign, combined with contralateral dissociated hypoesthesia mainly for pain and temperature.

Gillilan (1964) stressed that the intrinsic arterial patterns of the pons are more constant and predictable than those of the extrinsic arteries. The penetrating arteries supply primarily gray matter. He distinguished four major arterial zones:
1. Medial arteries, arising from the most medially situated extrinsic arteries
2. Paramedian arteries, branching from the short and long transverse arteries
3. Lateral arteries, branching from the transverse arteries and also from the AICA
4. Arteries of the dorsal zone, arising from the longest of the lateral arteries

According to Gillilan, the pattern of the lesion within the brain stem can be predicted from the distribution of the intrinsic arteries and does not bear any relation to the pattern of the superficial arteries. He therefore suggests that a clinical terminology must be based on the zone of the brain stem involved, and not on the name of the extrinsic artery. The term "inferior lateral pontine syndrome" is more exact than "anterior inferior cerebellar artery syndrome." For the same reason, the term "inferior posterior cerebellar artery syndrome" should be avoided. Instead the term "lateral medullary syndrome" should be used.

The pattern of vascular occlusion in lateral medullary infarctions was studied by Fisher et al. (1961). They analyzed 26 cases from the literature and 16 cases of their own and found that in 75% of the cases, the lateral medullary syndrome resulted from occlusion of the VA. Dissection studies showed that the lateral medullary region is supplied by five or six small arterial branches, which arise from the BA, the VA, the AICA, and the PICA. The variations in the vascularization of the area depend on the development of the AICA and the PICA and on the site of origin of the PICA. The special vulnerability to infarction of the lateral medullary region must be caused by the absence of collateral flow from one lateral medullary artery to the other.

In one case described by Fisher et al., a lateral medullary infarction was combined with inferolateral pontine infarction. In this case, in which an old occlusion of the left VA and a recent occlusion of the right VA existed, only a very thin and short AICA was present on the side of the lesion.

3.2 Analysis of Normal and Abnormal Angiograms

3.2.1 Materials and Methods

One hundred vertebral angiograms were analyzed of patients in whom no infratentorial lesions were present. In some of the older patients, atherosclerotic changes were present. Of the patients investigated, 44 were females and 56 were males. Ages ranged from 3 months to 76 years. Figure 23 shows the age distribution. The clinical diagnoses of these 100 patients are listed in Table 2. Further, 41 angiograms were analyzed of patients in whom an infratentorial tumor was present. This clinical material has been subdivided into three major groups: (a) cerebellopontine angle tumors (20 patients); (b) cerebellar tumors (15 patients); (c) brain stem tumors (six patients).

The ages of the 20 patients with cerebellopontine angle tumors ranged from 23 to 76 years; ten were males and ten were females. In 18 cases, CT scans

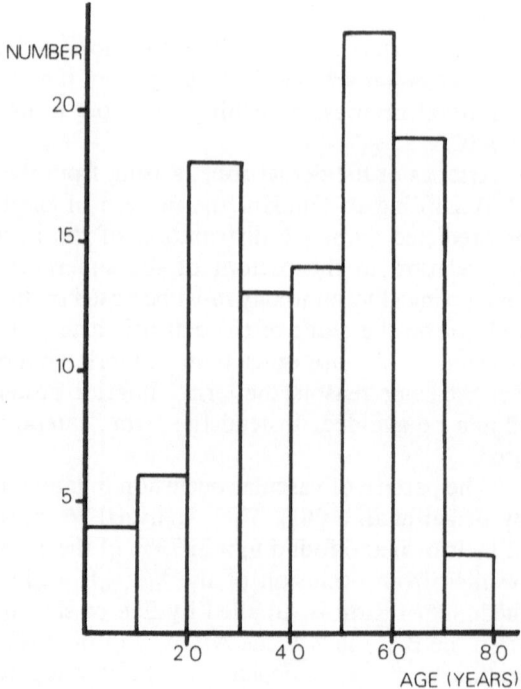

Fig. 23. Age distribution of 100 patients with normal vertebral angiograms

Table 2. Diagnoses of 100 patients with normal vertebral angiograms

Diagnosis	No. of patients
Suspected aneurysm or arteriovenous malformation	26
Supratentorial ischemic disturbance	46
Supratentorial tumor	17
Cranial trauma	3
Miscellaneous	8

were available. Fourteen patients had surgically confirmed acoustic neurinomas. In another two cases, this diagnosis was assumed likely on the basis of the clinical picture and the CT scan. No operations were performed in these two cases. In two cases, a glomus jugulare tumor was present. One patient suffered from a lipoma in the cerebellopontine angle. In one case, a metastasis from a hypernephroma was presumed present. This patient was also not operated.

The ages of the 15 patients with cerebellar tumors ranged from 1 to 72 years. Seven of them were children younger than 13 years. Eleven patients were males, and four were females. Diagnoses were: medulloblastoma (five patients), astrocytoma (three patients), hemangioblastoma (two patients), spongioblastoma (one patient), and metastasis (four patients). All diagnoses were verified operatively, with the exception of one of the metastatic tumors. All five cases of medulloblastoma, two cases of astrocytoma, and the one spongioblastoma were located in the midline. In the other seven cases, the tumor was located in one of the cerebellar hemispheres.

Four of the six patients with brain stem tumors were children aged 3–10 years. Only two patients were adults (48 and 66 years). Five patients were males, and one was female. Diagnosis of a pons glioma was made in all six cases, however, only two were operated on, and the diagnosis was confirmed histologically.

In addition to the 100 normal angiograms and the 41 angiograms of patients with posterior fossa tumors, nine angiograms of patients with transient or persistent ischemic lesions in the posterior cranial fossa were analyzed. These patients were 33–59 years old. Five of them were males and four were females.

In all cases, the Seldinger technique via the femoral artery was utilized. Using a thin-walled, 5.0-F catheter with a small terminal curve, the largest of the two vertebral arteries was catheterized. Fluoroscopy was used to ensure correct positioning of the catheter and the free flow of contrast material in a test injection. Subsequently, 8–10 ml Hexabrix® was injected. After injection, the catheter was immediately withdrawn.

Images were made with a biplane installation, allowing three films per second. As a rule, an AP projection of 20° angulation, cranial to the orbitomeatal line was chosen, bringing the petrous bones just above the roof of the orbit. In some cases, enlargement in the AP projection was used. Oblique projection was used additionally, in some cases, with cerebellopontine angle tumors. Subtraction films were always made. An empty roentgenogram was obtained as the first exposed film, prior to the contrast injection, in serial angiography.

3.2.2 Results

3.2.2.1 The AICA in 100 Normal Vertebral Angiograms

The configuration of the left and right AICA in the 100 angiograms reviewed was classified in the same way as described in Chap. 2. In each case, both sides were evaluated separately. Table 3 summarizes the results of this categorization.

No AICA Present. In 16 cases no AICA was seen on one side, and in two cases no AICA was seen bilaterally. In all cases in which the AICA was absent, the ipsilateral PICA was well developed.

Table 3. Frequency of occurrence of the main types of the anterior inferior cerebellar artery in 100 normal vertebral angiograms

	Bilateral	Unilateral	Total	Percentage
No AICA present	2	16	20	10
Type I	4	11	19	9.5
Type II	34	28	96	48
Type III	0	12	12	6
Type III A	3	17	23	11.5
Type IV	0	5	5	2.5
Type IV A	0	11	11	5.5
No categorization made	5	4	14	7

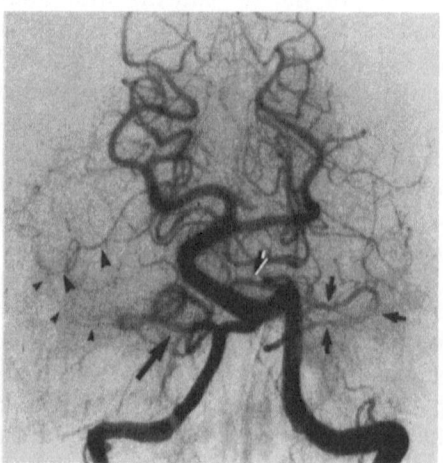

Fig. 24. Left vertebral angiogram, arterial phase, half-axial projection. The BA shows a right concave curve. There is elongation of all arteries. On the *right* side no AICA is present. The marginal branch (*large arrowheads*) of the SCA is well developed and takes over part of the area that is usually supplied by the lateral extension of the AICA. The right PICA (*large heavy arrow*) has a tortuous course. A hemispherical branch (*small arrowheads*) courses along the inferior aspect of the cerebellar hemisphere. The AICA on the *left* side is of the type IV configuration (*arrows*)

47

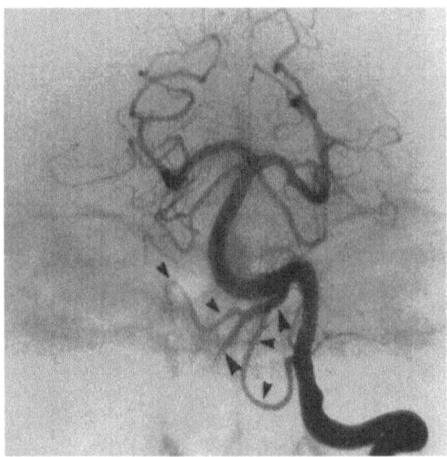

Fig. 25. Left vertebral angiogram, early arterial phase, half-axial projection. The VAs are asymmetrical, the left one being larger in caliber than the right. The latter is indicated by *large arrowheads*. On both sides, a well-developed PICA is present (*small arrowheads*). No AICAs are seen on either side

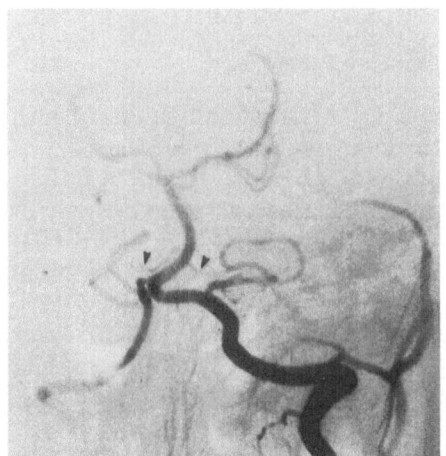

Fig. 26. Vertebral angiogram, arterial phase, half-axial projection. On both sides, a type I AICA is present (*arrowheads*). The AICAs are smaller in caliber than the PICAs and do not reach further laterally than the cerebellopontine angle

In Figs. 24 and 33, the AICA on the right side is missing. In Fig. 24, especially, the marginal branch of the SCA is well developed and anastomoses with branches of the PICA on the lateral aspect of the cerebellar hemisphere. Figure 25 shows a case in which AICAs are absent on both sides.

Type I AICA. In four patients, the AICA on both sides was of the type I configuration, and the artery was of small caliber; figure 26 shows an example. In 11 cases, the type I AICA was unilaterally present; figure 27 shows an example. Hemispherical branches of the SCA and the PICA are well developed and

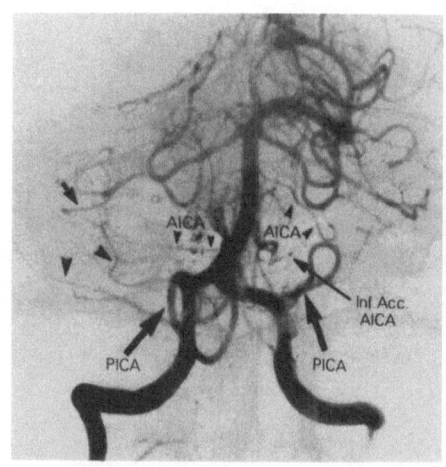

Fig. 27. Right vertebral angiogram, arterial phase, half-axial projection. There is no filling of the right posterior cerebral artery. The right AICA (*small arrowheads*) is thin and of the type I configuration. Branches of the SCA (*small heavy arrow*) and of the PICA (*large arrowheads*) are well developed and supply the lateral part of the cerebellar hemisphere. The left AICA arises from the distal half of the BA. A small inferior accessory AICA (*Inf. Acc. AICA*) arises from the BA at about the same level as the right AICA

anastomose with each other lateral to the short AICA. In all cases in which a type I AICA was present, the PICA on the same side was well developed.

Type II AICA. This configuration of the artery was seen in almost half of the cases investigated. Angiography in 34 patients revealed type II AICAs bilaterally. In 28 patients, the AICA on one side was of the type II configuration. Figure 28 shows the bilateral type II AICA. Both the arteries describe a loop in the vicinity of the internal auditory meatus. The type II AICA is not always of substantial caliber, as is shown in Fig. 29, in which both AICAs are thin, but show a considerable lateral extension. The right AICA forms a clearly defined meatal loop.

Usually, the AICA originates from the proximal half of the BA. It first travels in a caudolateral direction, then gradually changes course to a rostrolateral direction toward the internal auditory meatus, thus describing an inferior convex curve. In Fig. 28, this inferior convex loop of the proximal part of the AICA is visible in both the AP and lateral projections. A tortuous course of the proximal segment of the artery may sometimes occur, as illustrated in Fig. 30. This case also illustrates the advantage of the early filling phase in recognizing the exact course of the proximal part of the AICA, which in a later stage is partially obscured by the branches of the PICA. A more or less straight laterally directed course, as in Fig. 31, is exceptional.

In 16 of 96 type II AICAs, no meatal loop was visible. In the other 80 cases, the angiographic appearance of the loop in the vicinity of the internal auditory meatus was variable, as illustrated by Fig. 69. Branching of the type II AICA usually occurs distal to the meatal loop. A small caudomedial branch that originates from the proximal portion of the artery is an exception; figure 32

49

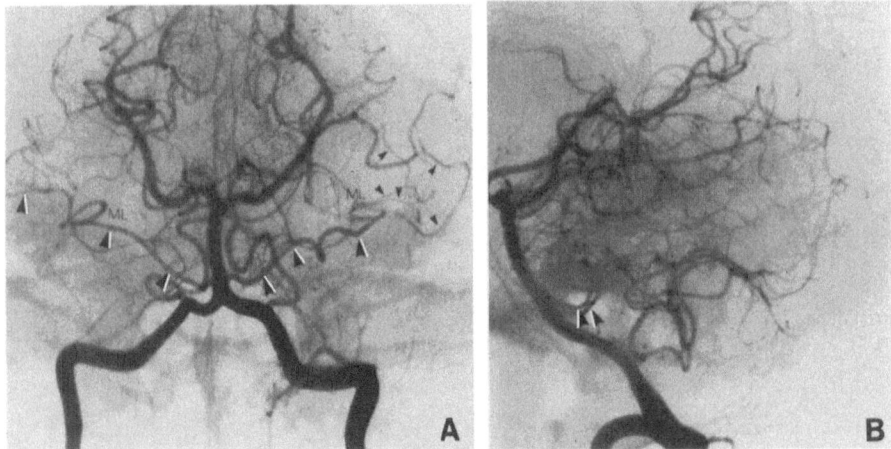

Fig. 28A, B. Left vertebral angiogram, arterial phase. A Half-axial projection. Both AICAs (*large arrowheads*) display meatal loops (*ML*) and are of the type II configuration. Especially on the *left* side, the different hemispherical branches (*small arrowheads*) are clearly shown. The calibers of the AICAs are approximately equal to the calibers of the PICAs. B Lateral projection. The premeatal segments (*arrowheads*) of the AICAs are symmetrically developed and show inferior convex curves. The more distal part of the two AICAs is difficult to distinguish due to overprojection

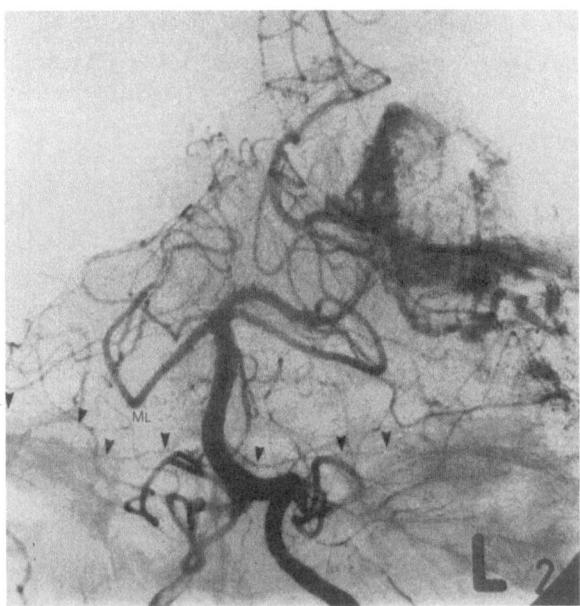

Fig. 29. Left vertebral angiogram, arterial phase, half-axial projection. Both the right and left AICAs are of small caliber (*arrowheads*). A meatal loop (*ML*) is present on the *right* side. (Abnormal tumor vessels of a left-sided occipital lobe tumor are present)

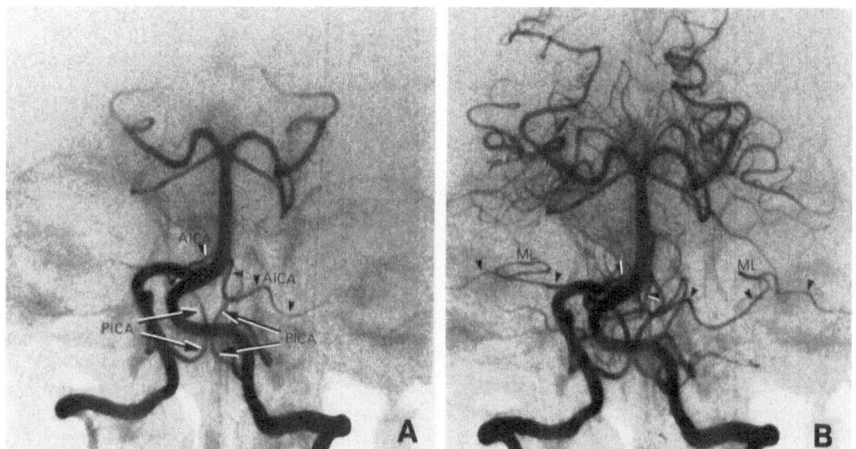

Fig. 30 A, B. Left vertebral angiogram, half-axial projection. **A** Early arterial phase. Only the proximal parts of both the PICAs and AICAs are shown. The proximal segment of the left AICA shows a tortuous course. **B** Arterial phase (1 s later than A). Both AICAs are of the type II configuration and in this phase, the meatal loops are visualized (*ML*). Because of overprojection of PICA branches, it is difficult to identify the exact course of the proximal part of the left AICA

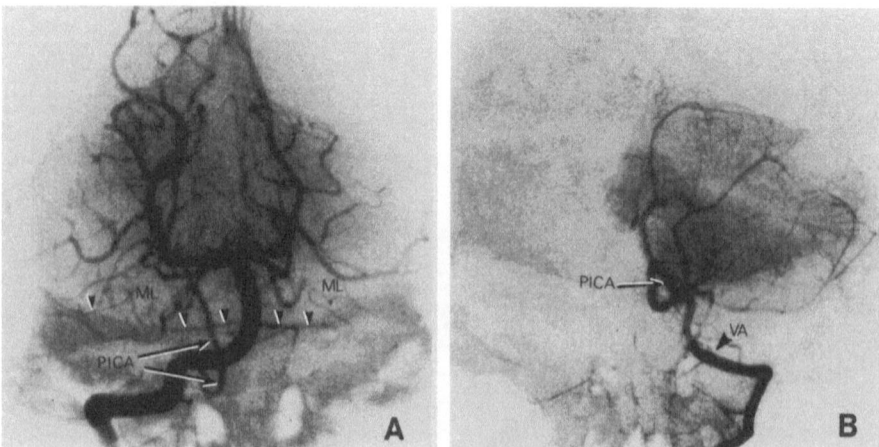

Fig. 31. A Right vertebral angiogram, arterial phase, half-axial projection. There is no filling of the left VA and the left PICA. The right and left AICAs arise from the same level of the BA. On both sides, the proximal segment shows a straight, laterally directed course. **B** Left vertebral angiogram, arterial phase, half-axial projection. The left VA does not unite with the right VA, and injection of the left VA does not achieve filling of the BA and its branches. There is only filling of the area of the PICA

shows an example. The right AICA bifurcates before reaching the internal auditory meatus, and one branch turns in a caudomedial direction toward the brain stem.

Distal to the internal auditory meatus, the AICA usually divides into several hemispherical branches; this is illustrated in Fig. 28. In this case, three hemi-

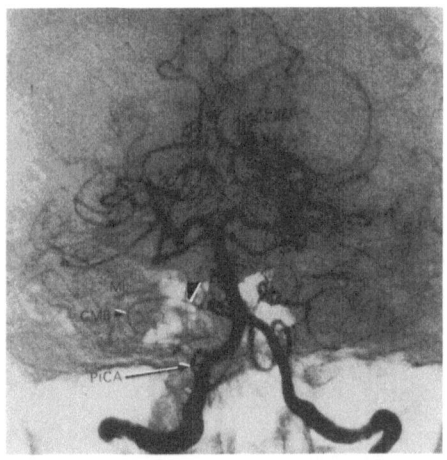

Fig. 32. Left vertebral angiogram, arterial phase, half-axial projection. The right AICA (*arrowhead*) is of small caliber. The artery bifurcates into a caudomedial branch (*CMB*) and a rostrolateral branch. The meatal loop (*ML*) is part of the latter branch

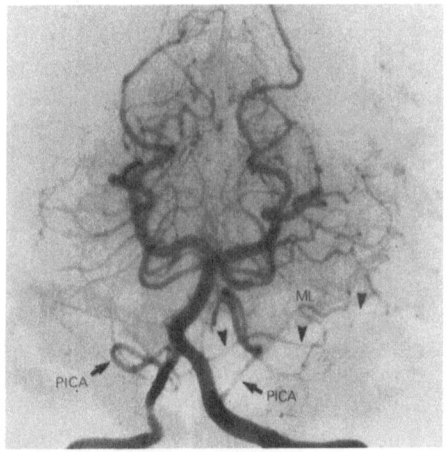

Fig. 33. Left vertebral angiogram, arterial phase, half-axial projection. On the *right* side, only a PICA is present, and no AICA is seen. On the left side, the PICA and the AICA (*arrowheads*) both originate from the VA. The AICA is of the type II configuration

spherical branches supply the superior and inferior semilunar lobules of the cerebellum.

In the majority of cases in which a type II AICA was present, the caliber of the AICA was smaller than the caliber of the PICA (in 72 cases). The calibers of the AICA and the PICA were about equal in 20 cases, and in four cases only the type II AICA was larger in caliber than the ipsilateral PICA. The PICA was not absent in any of the 96 cases of a type II AICA.

In one case (Fig. 33), a type II AICA originated from the VA. The ipsilateral PICA originated from the same VA in this case.

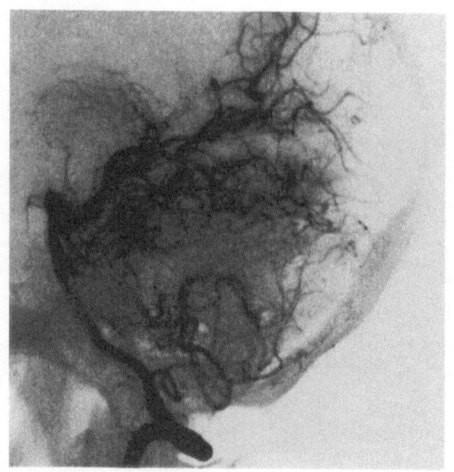

Fig. 34. Left vertebral angiogram, arterial phase, lateral projection. The proximal parts of the AICAs are obscured by the petrous bones

In the lateral view, it was often not possible to visualize the different parts of the AICA as clearly as in the AP view. In Fig. 34, the proximal parts of the AICAs are not visible because of overprojection of the petrous bones. In Fig. 35, the meatal loops of the right and left AICA are clearly visible in the AP view. In the lateral projection, however, it is difficult to tell both AICAs apart.

The distal branches of the AICA may be obscured by either the branches of the contralateral AICA or by the branches of the PICAs. An example is shown in Fig. 36. In the case shown in Fig. 37, asymmetrical AICAs are present, the right one being larger in caliber than the left. The right AICA divides into two main branches, one ascending and one descending. In the lateral projection, the distal parts of these branches are obscured by the branches of the PICAs. However, the proximal segments and both meatal loops can be recognized quite well in the lateral projection. The meatal loops in this case have approximately the same M-shaped configuration in the lateral projection, as is shown in Figs. 28 and 36.

In 18 of 80 cases in which a meatal loop was visible in the AP view, the lateral view showed a recognizable M-shaped meatal loop. In the other cases, either another configuration of the meatal loop was present, or the exact configuration could not be determined (as in Fig. 35).

Type III and Type III A AICAs. In 12 cases, a unilateral type III AICA was present. A bilateral type III AICA was not observed. The type III A AICA was seen 17 times unilaterally and three times bilaterally. Figure 38 shows a case in which a type III AICA is present. The diameter of the main stem of the AICA is about equal to that of the ipsilateral PICA. The latter artery does not give off hemispherical branches but supplies only the caudal part of the vermis cerebelli. In this case, the caudomedial branch of the AICA takes over the hemispherical supply of the PICA. As shown in Fig. 39, however,

53

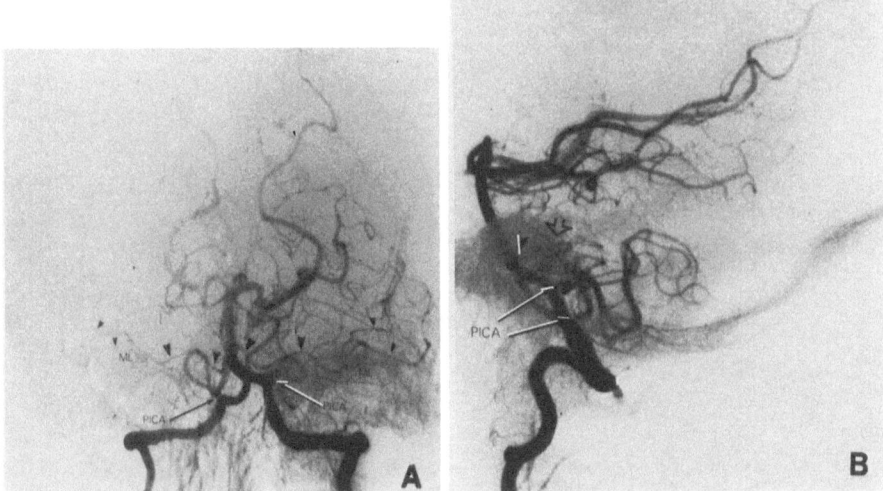

Fig. 35A, B. Left vertebral angiogram, arterial phase. **A** Half-axial projection. On both sides, a type II AICA (*large arrowheads*) is present, and both AICAs form a meatal loop (*ML*). The hemispherical branches of the AICAs are indicated by *small arrowheads*. **B** Lateral projection. The proximal segments of both AICAs (*large arrowheads*) can easily be recognized. It is difficult to distinguish between the more distal parts of the arteries (*open arrow*)

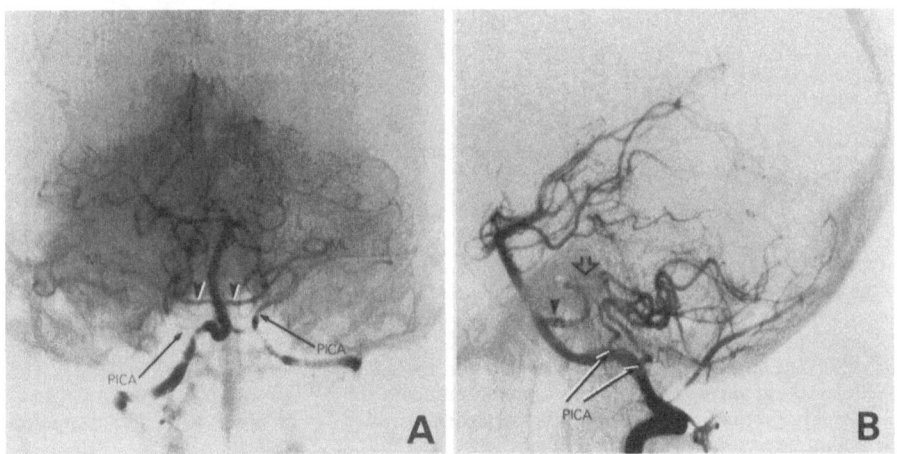

Fig. 36A, B. Right vertebral angiogram, arterial phase. **A** Half-axial projection. Both the right and left AICAs (*arrowheads*) and the right and left PICAs are more or less symmetrically developed. A meatal loop (*ML*) is present on both sides. **B** Lateral projection. The proximal segments of the right and left AICAs (*arrowheads*) show overprojection. The meatal loop (*open arrow*) can be identified, but the distal segments of both AICAs cannot be recognized because of overprojection

this is not always the case. The caudomedial branch of the AICA in this case supplies part of the cerebellar hemisphere, but a branch of the PICA also contributes to the vascular supply of the cerebellar hemisphere. This case also illustrates that a meatal loop in the type III AICA is not always present.

54

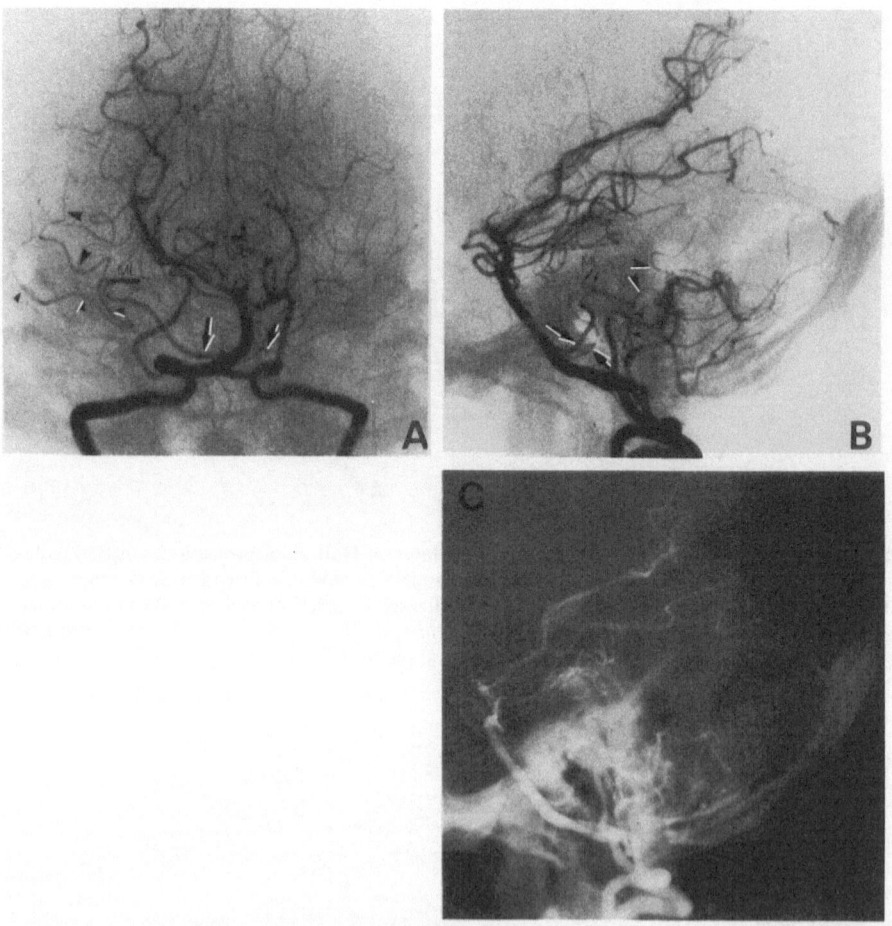

Fig. 37 A–C. Left vertebral angiogram, arterial phase. **A** Towne projection. On both sides, a type II AICA is present (*arrows*), but the right one is larger in caliber than the left. Distal to the meatal loop (*ML*), the right AICA divides into an ascending and a descending branch (*small arrowheads*). The left AICA also forms a loop near the internal acoustic meatus (*ML*). **B** Lateral projection. The right and left AICAs (*arrows*) are easy to distinguish because of the difference in caliber. The same is true for the right and left meatal loops (*ML*). The ascending (*large arrowheads*) and descending branches (*small arrowheads*) can be recognized. **C** Lateral projection, nonsubtracted film. The details which can be observed in the subtraction (**B**) cannot be seen in this nonsubtracted film

Figure 40 shows the III A configuration on the left side. In both the AP and lateral projections, the caudomedial branch is well visible. This branch of the AICA provides those areas of the left cerebellar hemisphere that are usually supplied by the PICA. The rostrolateral branch is only of small size, but a meatal loop is present. The vascular configuration on the right side in this case is exactly the reverse – no AICA is present.

In Fig. 41, the rostrolateral branch and the caudomedial branch on the right side are about equally developed.

In Fig. 42, an exceptional case of a so-called megadolichobasilar artery is shown. On both sides, a type III A AICA is present.

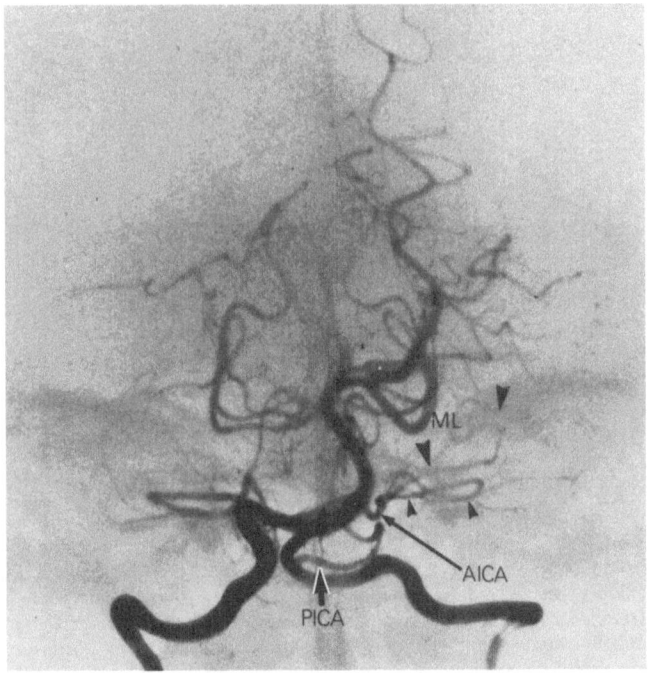

Fig. 38. Right vertebral angiogram, arterial phase, half-axial projection. The BA shows a right concave curve. On the *left* side, the AICA and the PICA are about equal in diameter. The AICA divides into a rostrolateral branch (*large arrowheads*) and a caudomedial branch (*small arrowheads*). The PICA supplies only a part of the cerebellar vermis and does not contribute to the supply of the left cerebellar hemisphere

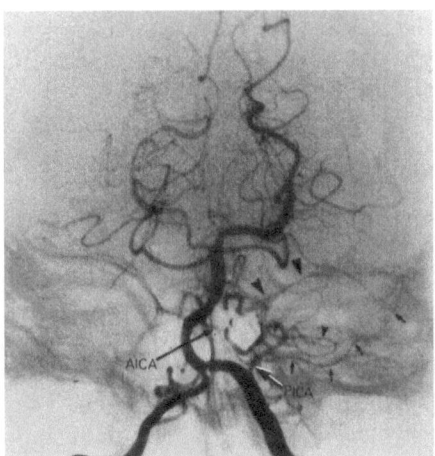

Fig. 39. Left vertebral angiogram, arterial phase, half-axial projection. On the *left* side, there is a type III AICA. The rostrolateral branch is indicated by *large arrowheads* and the caudomedial branch by *small arrowheads*. Despite the large area of supply of the AICA, on the *left* side there is a well-developed PICA, which sends branches (*small arrows*) to the inferior aspect of the cerebellar hemisphere

56

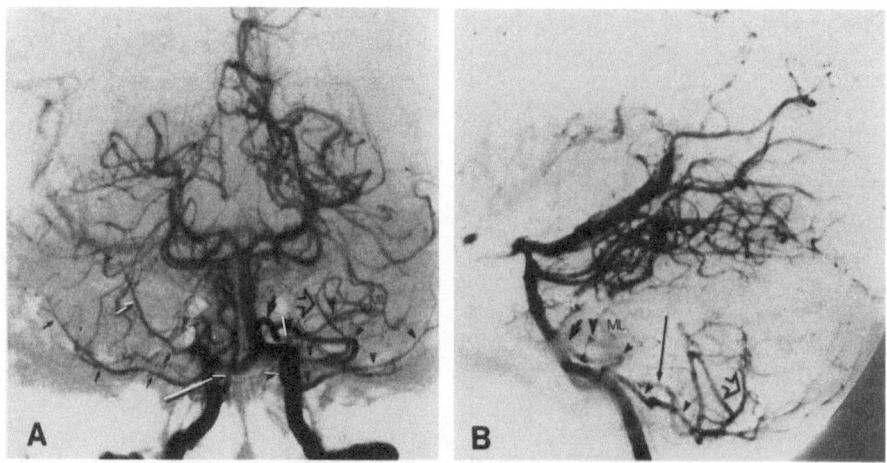

Fig. 40 A, B. Vertebral angiogram, arterial phase. **A** Half-axial projection. On the *right* side, no AICA is present. The right PICA (*long arrow*) sends large branches to the right cerebellar hemisphere (*small arrows*). On the *left* side, the PICA is absent, and a large AICA is present. The proximal part of the artery (*heavy arrow*) shows a tortuous course. The relatively thin rostrolateral branch is indicated by *large arrowheads*. This branch forms a meatal loop (*ML*). The caudomedial branch (*small arrowheads*) is much larger in diameter and curls around the inferior aspect of the cerebellar hemisphere. A small branch runs upwards on the posterior aspect of the cerebellar hemisphere (*open arrow*). **B** Lateral projection. Symbols used are the same as in **A**. The course of the caudomedial branch along the inferior aspect of the cerebellar hemisphere is clearer in this projection than in the half-axial view. The rostrolateral branch is foreshortened

Of the 23 cases in which type III A AICAs were present, the caudomedial branch and the rostrolateral branch were in eight cases about equal in size. In 12 cases, the caudomedial branch was larger than the rostrolateral branch, and in three cases, the rostrolateral branch was larger than the caudomedial branch. In the 11 cases of type III AICAs, these figures were about the same: The branches were equal in size in six cases, the caudomedial branch was larger in five cases, and the rostrolateral branch was larger in one.

Type IV and Type IV A AICAs. In five cases, a unilateral type IV AICA was present. An example is shown in Fig. 43. There were no cases with bilateral type IV AICAs. In two cases in which a type IV AICA was present, the AICA was smaller than the ipsilateral PICA, and in three cases the AICA was larger than the ipsilateral PICA. In 11 cases, a unilateral type IV A AICA was seen. There were no cases with bilateral type IV A AICAs. Figure 44 shows a case with the type III AICA on the right and the type IV AICA on the left. The marginal branch of the left SCA is strongly developed. Both the right and the left PICA are small. Figure 45 shows a case in which a type IV A AICA is present on the left side. The artery divides into two main branches: a branch to the caudal part of the cerebellar vermis and a branch to the cerebellar hemisphere. The latter branch divides into several smaller hemispherical branches. The marginal branch is strongly developed.

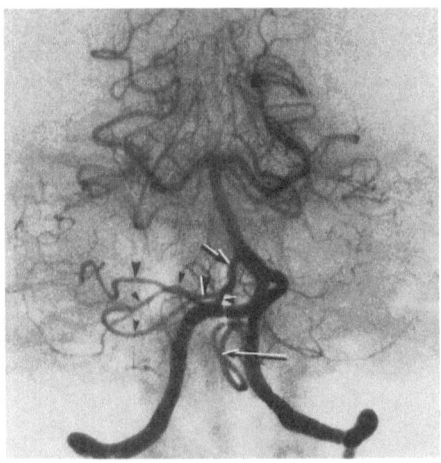

Fig. 41. Left vertebral angiogram, arterial phase, half-axial projection. On the *right* side, a type III A AICA is present. The most proximal part of the artery is indicated by the *heavy arrow*. The artery divides into two main branches – a rostrolateral branch (*large arrowheads*) and a caudomedial branch (*small arrowheads*). Only on the *left* side is a PICA present (*long arrow*). No AICA is seen on this side

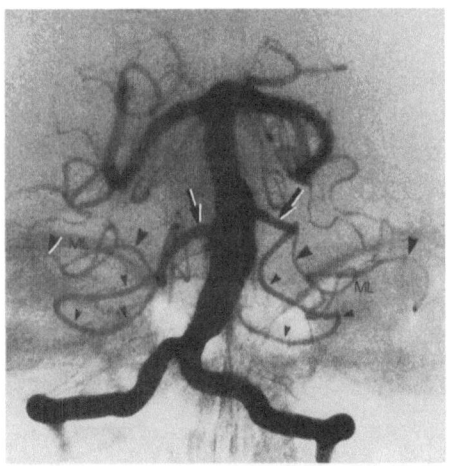

Fig. 42. Left vertebral angiogram, arterial phase, straight AP projection. There is a megadolichobasilar artery. No PICAs are present on either the *right* or *left* sides. The AICAs are more or less symmetrical and are both of the type III configuration. Both AICAs divide into a rostrolateral (*large arrowheads*) and a caudomedial (*small arrowheads*) branch. Complicated meatal loops (*ML*) are part of the rostrolateral branch on both sides

No Categorization Possible. In 14 cases, no categorization according to the classification described could be made. In six instances (i.e., in three patients), this was due to overprojection of the ipsilateral PICA in the Towne or half axial projection, which made it impossible to recognize the exact configuration of the AICA. In seven cases, no classification was made because there was multiplication of the AICA, and both arteries were of about equal size. Therefore,

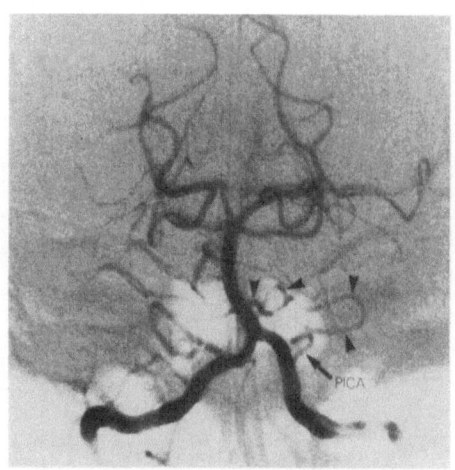

Fig. 43. Right vertebral angiogram, arterial phase, half-axial projection. A type IV A AICA is present on the *left* side (*arrowheads*). The PICA is smaller in caliber than the AICA

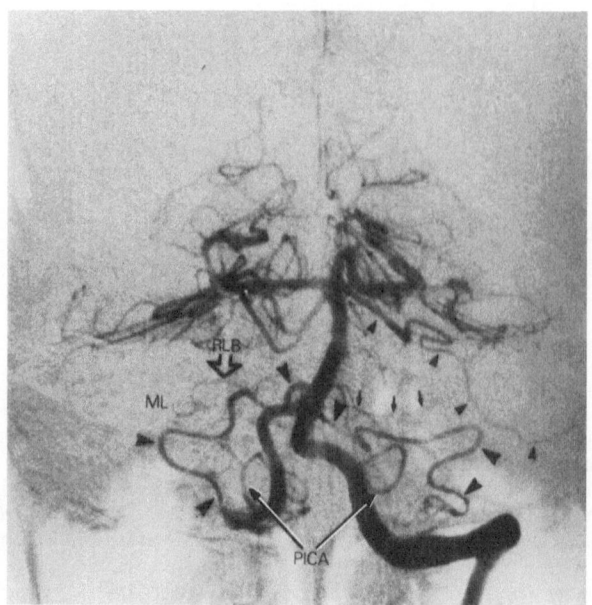

Fig. 44. Left vertebral angiogram, arterial phase, straight AP projection. On the *right* side, there is only a small PICA. The AICA is much larger and is of the type III configuration (*large arrowheads*). The rostrolateral branch is small. The meatal loop (*ML*) is part of it. On the *left* side, the PICA is also smaller than the AICA. The left AICA (*large arrowheads*), however, is of the type IV configuration, because no rostrolateral branch is present. The marginal branch (*small arrowheads*) is well-developed and runs in the lateral part of the horizontal fissure. A small superior accessory AICA is seen on the *left* side (*small arrows*)

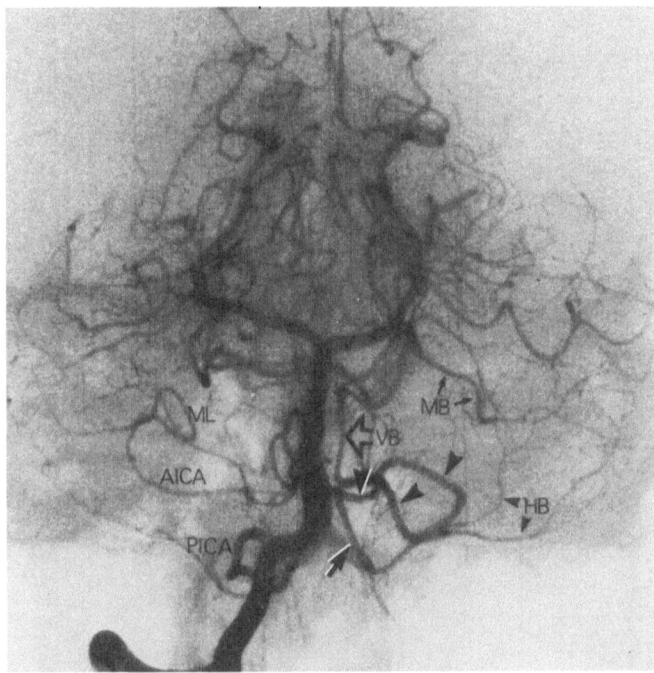

Fig. 45. Right vertebral angiogram, arterial phase, half-axial projection. The right AICA is of the type II configuration and forms a meatal loop (*ML*). On the *left* side, no PICA is present. The left AICA is indicated by *large arrowheads*. After having described a large loop over the cerebellar hemisphere, the artery divides into a vermian branch (*VB*) and a branch (*heavy arrow*) which further divides into hemispherical branches (*HB*). The marginal branch (*MB*) of the SCA is well developed and extends laterally to the most lateral part of the horizontal fissure

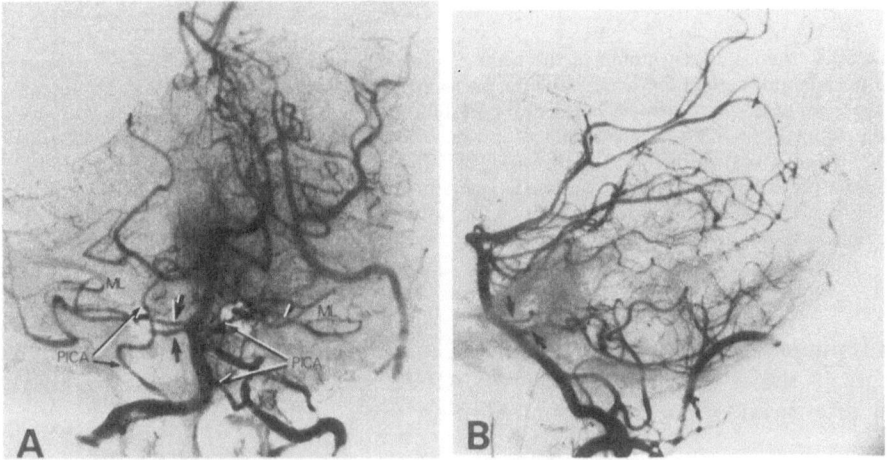

Fig. 46 A, B. Left vertebral angiogram, arterial phase. **A** Half-axial projection. Two AICAs are present on the *right* (*heavy arrows*). Both are of the type II configuration, and one of the arteries forms a meatal loop (*ML*). The right PICA is of small caliber. On the *left* side, one AICA is present (*arrowheads*), which is of the type II configuration. **B** Lateral projection. The proximal parts of both right AICAs and of the one left AICA can be recognized

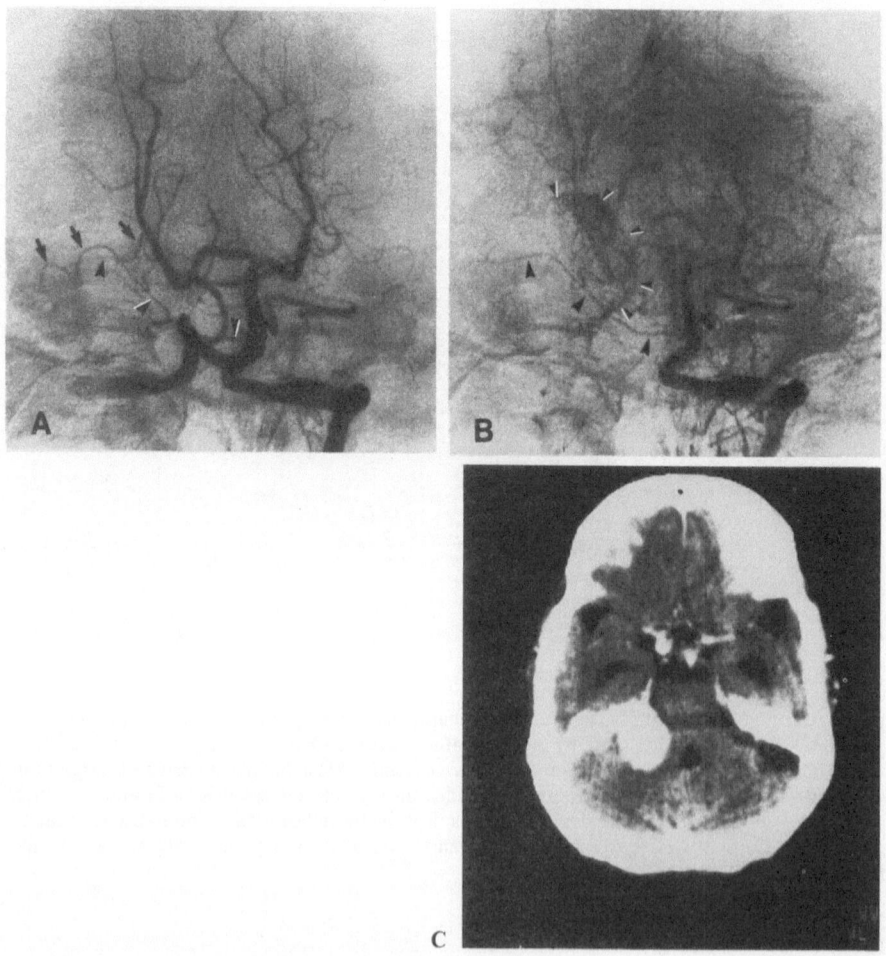

Fig. 47 A–C. Acoustic neurinoma in the right cerebellopontine angle; female, aged 75 years. **A** Left vertebral angiogram, arterial phase, half-axial projection. A type II AICA is present on the *right* side (*large arrowheads*). The part of the artery which is located distal to the meatal loop (*ML*) is lifted and displaced posteriorly. There is apparently no displacement of the marginal branch of the SCA (*arrows*). **B** Left vertebral angiogram, late arterial phase, half-axial projection. Staining of the medial part of the tumor (*small arrowheads*), is clearly visible. **C** CT scan after contrast injection. A contrast-enhanced, well-circumscribed tumor is present in the right cerebellopontine angle

it was not possible to nominate either one of the arteries as an accessory AICA. Figure 46 shows a case in which two AICAs are present on the right side; both arteries are of the type II configuration.

3.2.2.2 The AICA in Cerebellopontine Angle Tumors

Tumor stains were present in 4 of 20 cases investigated. Two of these cases were acoustic neurinomas, of which Fig. 47 shows an example. It is important

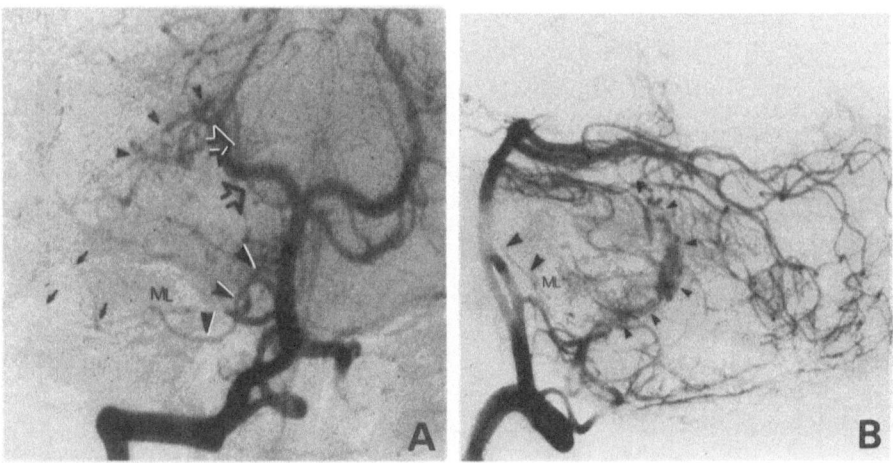

Fig. 48. Acoustic neurinoma in the right cerebellopontine angle; female, aged 60 years. Right vertebral angiogram, arterial phase. **A** Half-axial projection; **B** Lateral projection. A large hypervascular tumor (*small arrowheads*) elevates the proximal portion of the SCA (*open arrow*). There is a type II AICA on the *right*, of which both the proximal part (*large arrowheads*) and the meatal loop (ML) are inferiorly displaced. Distal branches of the AICA (*small arrows*) supply a large part of the cerebellar hemisphere

to note that especially the medial part of the tumor is stained after contrast injection into the VA. In Fig. 48, tumor staining is also present. In three cases, no AICA could be identified angiographically. In two of these cases, the AICA is probably absent, but in one case (Fig. 49), in which a very large acoustic neurinoma was present, it is possibile that the AICA is severely distorted and cannot therefore be recognized. Small tortuous vessels are present in this case.

A type I AICA on the side of the lesion was seen in three patients, and Fig. 50 shows one of these cases. Apparently, there is no significant displacement of this vessel, which is in accordance with the finding during the operation that only a small acoustic neurinoma was present. No clear displacement was observed in the other two cases of type I AICAs.

Type II AICAs were present in ten instances. A superior shift of the artery was seen in seven cases. In the other three cases, inferior displacement was present. In Fig. 47, a superior shift of the artery is shown. In Figs. 51 and 52, only a portion of the AICA shows a superior displacement. In these cases, there is probably also a posterior displacement of the artery. Superior and posterior displacement of the type II AICA is also visible in Fig. 53. On the original (not subtracted) film, especially, it becomes clear that the AICA is pressed away from the petrous bone. Figures 54 and 55 show examples of an upward displacement of a type II AICA in combination with a medial shift. In three cases, inferior displacement of a type II AICA was seen. Figures 56 and 47 show examples of inferiorly displaced type II AICAs. This type of displacement is well illustrated in the lateral projection.

No type III AICAs were present in the material reviewed. In four cases, a type IV or type IV A AICA was seen. Figure 57 shows a case in which the proximal part of the type IV AICA is stretched. The part of the artery which courses near the meatus acusticus internus is elevated. Distal to this, the artery

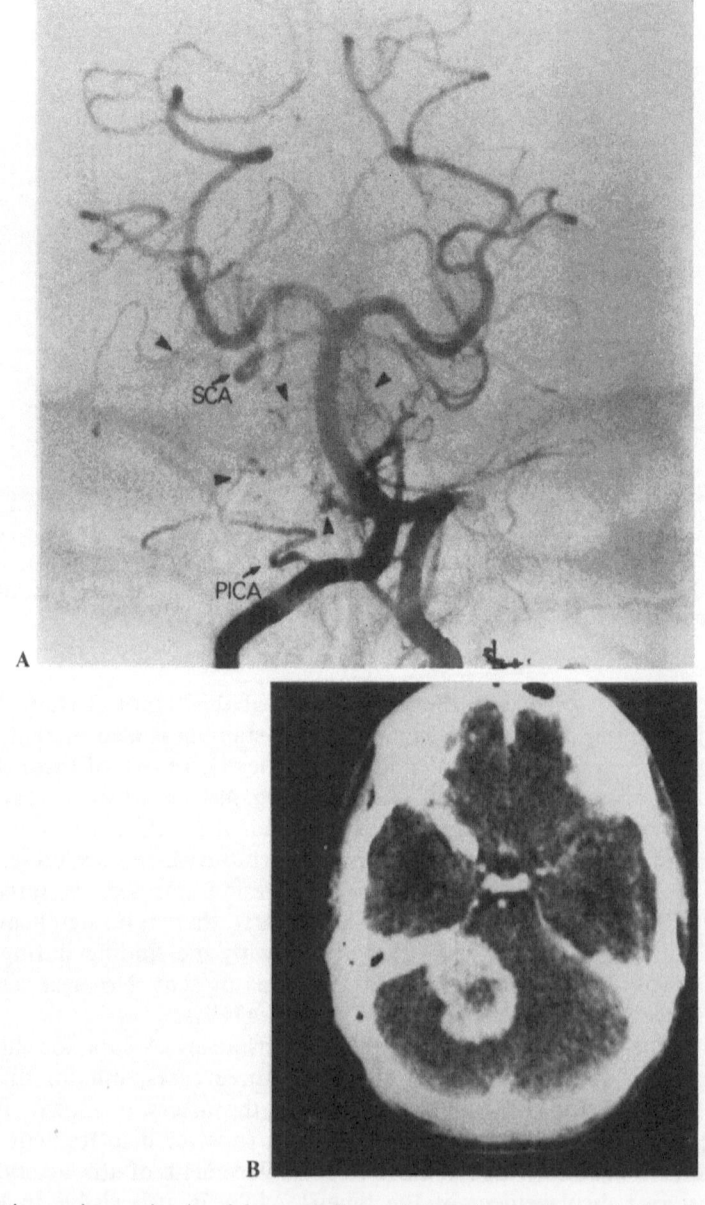

Fig. 49 A, B. Acoustic neurinoma in the right cerebellopontine angle; male, aged 28 years. **A** Left vertebral angiogram, arterial phase, half-axial projection. The right SCA is medially displaced and makes a sharper curve than the left. Only a small PICA is present on the *right* side, a finding which is often associated with the presence of a large AICA. However, no AICA is observed on the *right* side. Small irregular tumor vessels can be seen, which cross the midline and extend to the *left* side. **B** CT scan after contrast injection. A large, contrast-enhanced tumor is present in the left cerebellar hemisphere

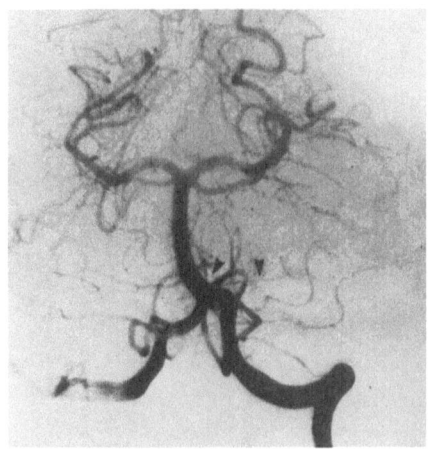

Fig. 50. Small acoustic neurinoma in the left cerebellopontine angle; male, aged 54 years. Left vertebral angiogram, arterial phase, half-axial projection. A type I AICA is present on the *left* side (*arrow-heads*). The artery shows an inferiorly convex curve. There is no obvious displacement of the vessel

A   B

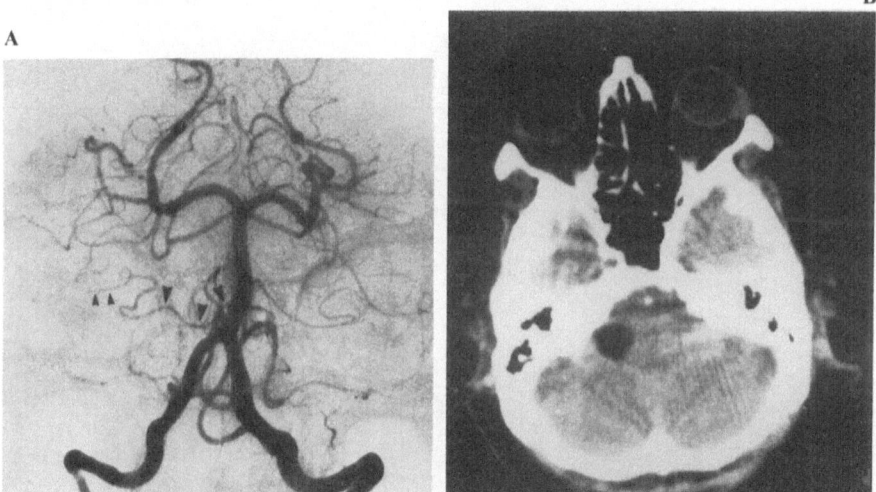

Fig. 51 A, B. Lipoma in the right cerebellopontine angle; male, aged 62 years. A Right vertebral angiogram, arterial phase, half-axial projection. On both sides, a type II AICA is present. The right AICA (*large arrowheads*) shows a normal course in its proximal portion. The distal part of the artery (*small arrowheads*) is somewhat displaced into a superior and posterior direction. B CT scan. There is a well-circumscribed, low-density lesion in the right cerebellopontine angle

turns back in a medial direction. In the case presented in Fig. 58 the type IV AICA obviously supplies the larger part of the inferior aspect of the right cerebellar hemisphere, and the ipsilateral PICA is small.

3.2.2.3 The AICA in Cerebellar Tumors

In the eight cases of midline tumors, there was no severe displacement of the AICAs. Figure 59 shows a case of medulloblastoma. On the right side, a type II

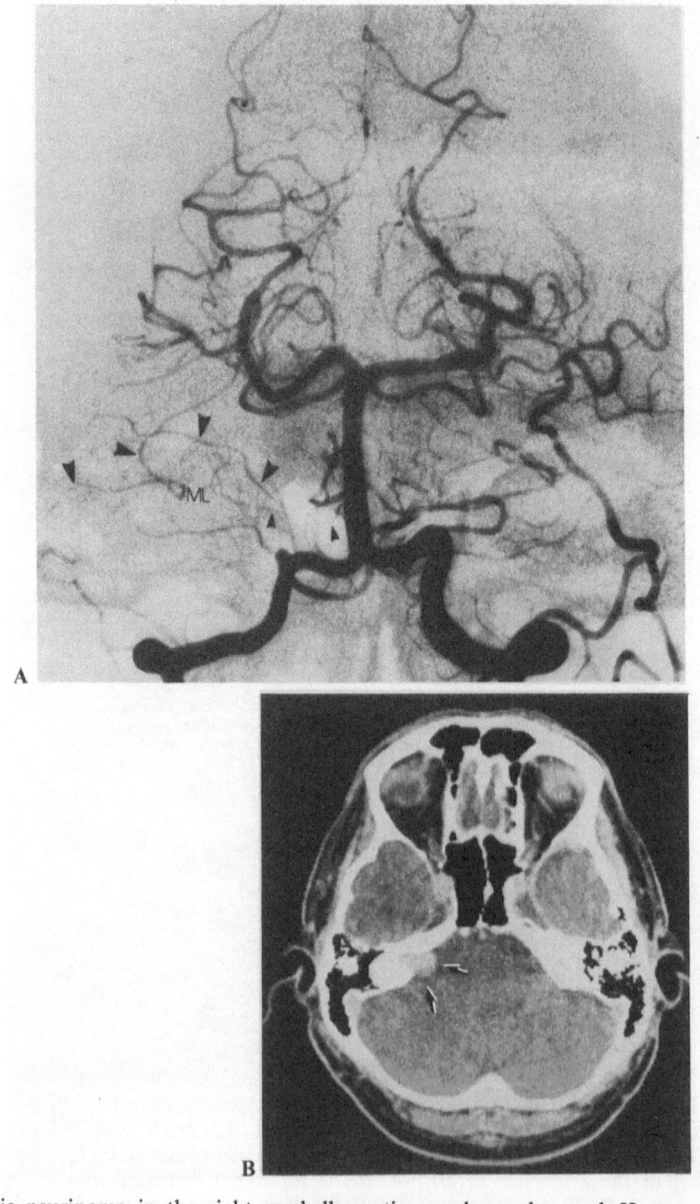

Fig. 52A, B. Acoustic neurinoma in the right cerebellopontine angle; male, aged 52 years. **A** Right vertebral angiogram, arterial phase, half-axial projection. The right type II AICA is somewhat elevated in its premeatal portion (*large arrowheads*). There is an inferior accessory AICA running in the direction of the internal auditory meatus (*small arrowheads*). **B** CT scan after contrast injection. A small, contrast-enhanced, partly intracanalicular, acoustic neurinoma is present in the right cerebellopontine angle (*arrows*)

AICA and on the left side a type IV A AICA are seen. In the AP view, no displacement of either artery can be seen. In the lateral view, arterial branches show a stretched course, surrounding the tumor. No tumor vessels are observed.

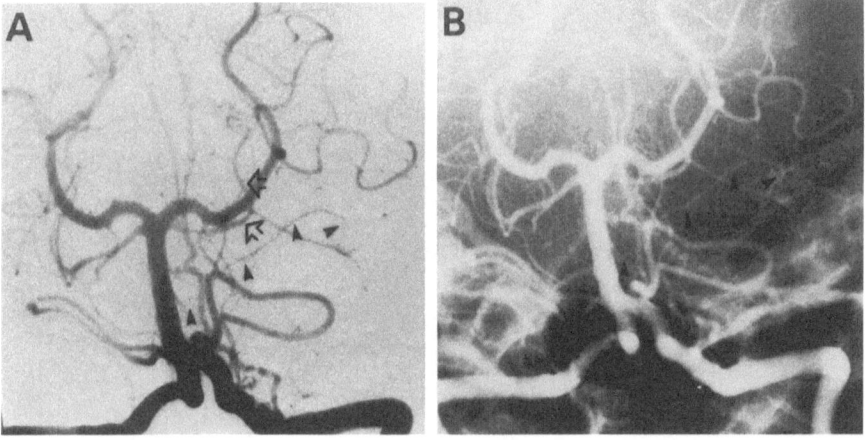

Fig. 53A, B. Acoustic neurinoma in the left cerebellopontine angle; female, aged 53 years. Left vertebral angiogram, arterial phase, Towne projection. **A** Subtracted film. There is elevation of the proximal part of the SCA (*open arrows*). The AICA is of very small caliber (*arrowheads*) and has an arched course. **B** Non-subtracted film. From this original film, it is clear that the proximal part of the artery is pressed away from the petrous bone (*arrowheads*)

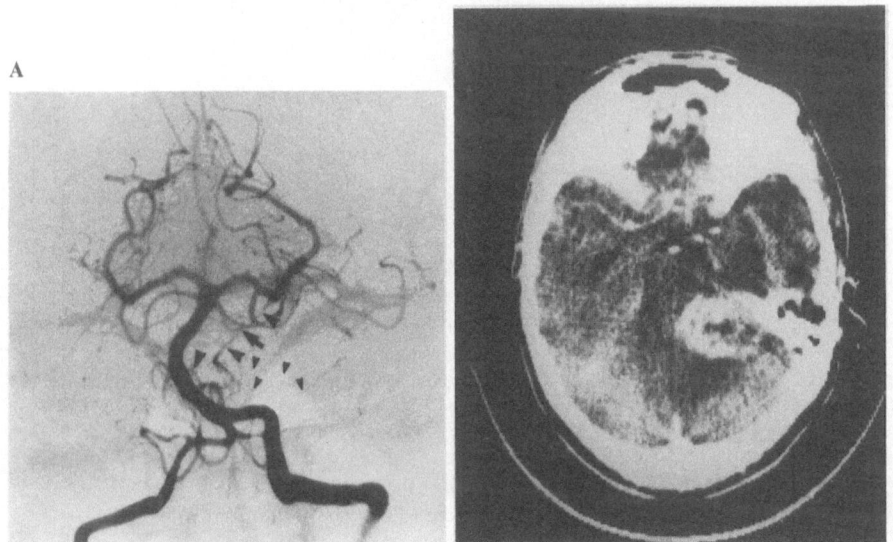

Fig. 54A, B. Acoustic neurinoma in the left cerebellopontine angle; female, aged 58 years. **A** Left vertebral angiogram, arterial phase, half-axial projection. The BA shows a left concave curve. The AICA is superiorly and medially displaced (*large arrowheads*). The meatal loop (*arrow*) projects behind the SCA. Two small stretched arteries (*small arrowheads*) originate from the AICA and run in a lateral direction toward the internal auditory meatus. **B** CT scan after contrast injection shows a large contrast-enhanced mass in the left cerebellopontine angle

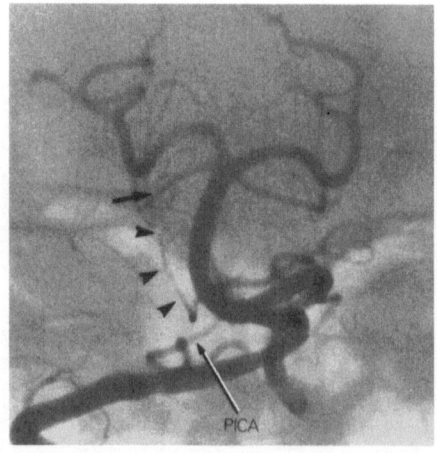

Fig. 55. Right cerebellopontine angle mass; male, aged 48 years. Right vertebral angiogram, early arterial phase, half-axial projection. Marked tortuosity of the VAs and the BA exists. The right AICA is elevated and displaced toward the midline (*arrowheads*). The meatal loop projects behind the SCA (*heavy arrow*)

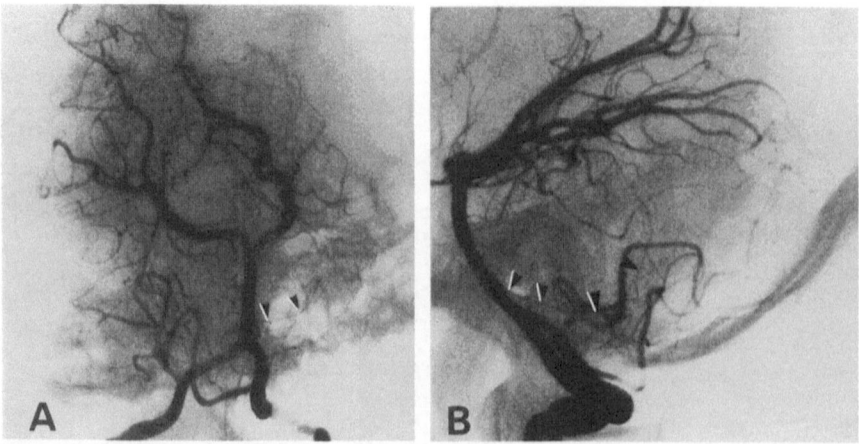

Fig. 56 A, B. Acoustic neurinoma in the left cerebellopontine angle; male, aged 37 years. **A** Left vertebral angiogram, arterial phase, oblique projection. The left AICA (*arrowheads*) is small. In this view, the artery is projected free from surrounding vessels, which is not the case in the AP projection. Possible displacement of the artery cannot be judged in this view. **B** Left vertebral angiogram, arterial phase, lateral projection. The course of the left AICA is stretched, and the artery is displaced inferiorly by the tumor (*arrowheads*)

In one of seven cases of cerebellar hemisphere tumor, the AICA was absent on the side of the tumor. Type II AICAs were present in five cases. No severe displacements of these type II AICAs could be observed in the AP and lateral views. Figure 60 shows a case in which a type II AICA is present on the side

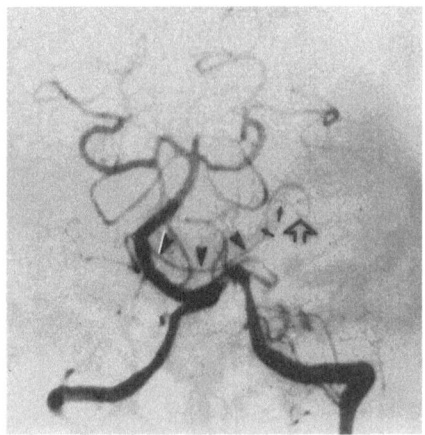

Fig. 57. Glomus jugulare tumor extending into the left cerebellopontine angle; female, aged 60 years. Left vertebral angiogram, arterial phase, transfacial projection. There is displacement of the BA to the *right*. The course of the proximal part of the type IV AICA is stretched (*arrowheads*), and more distally the artery is elevated (*open arrow*). At this point, the artery turns in a medial direction (*small arrows*).

of the tumor. The artery contributes to the staining of a large metastasis in the cerebellar hemisphere.

Figure 61 shows the vertebral angiogram in a case of hemangioblastoma in the right cerebellar hemisphere. A type IV A AICA is present here. Displacement of the artery is difficult to confirm in the AP view. In the lateral view, the artery shows a stretched course and is antero-inferiorly displaced. The tonsillar loop of the left PICA is located below the foramen magnum as a sign of tonsillar herniation. As the right PICA is missing, this sign is absent on the right side.

3.2.2.4 The AICA in Brain Stem Tumors

In the six cases of brain stem tumors, three cases showed no definite abnormality in the course of the AICAs. Figure 62 shows a case in which a type II AICA is present on both sides. The proximal portions of both AICAs show a stretched course and some inferior displacement. The latter is particularly well seen in the lateral projection. A similar course was observed in two further cases.

3.2.2.5 The AICA in Ischemic Lesions

Nine angiograms of patients with transient or persistent ischemic lesions in the posterior cranial fossa were reviewed. The clinical diagnoses and angiographic findings are summarized in Table 4. In three cases [A, E, and H (Fig. 63)], no stenosis or occlusion of major arteries could be demonstrated. Lesions of

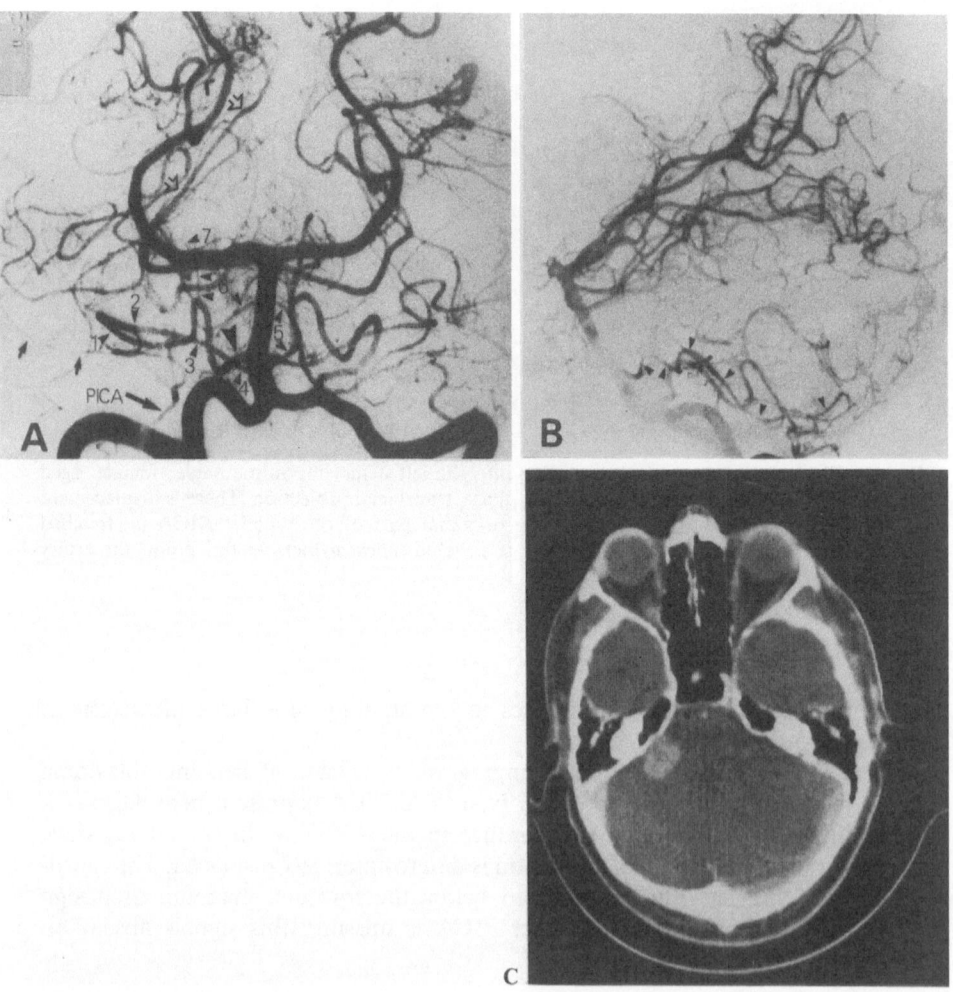

Fig. 58 A–C. Acoustic neurinoma in the right cerebellopontine angle; male, aged 38 years.
A Left vertebral angiogram, arterial phase, half-axial projection. The right SCA is medially
displaced (*open arrow*). The right PICA, which is small (*large arrow*), sends some branches
to the inferior aspect of the cerebellar hemisphere (*small arrows*). A type IV AICA is present
on the *right* side. A *large arrowhead* points to the origin of the artery from the BA. *Small
arrowheads* indicate the course of the artery and are numbered *1–7*, the higher numerals being
more distal. Immediately after its origin, the AICA describes an inferior convex loop. Between
1 and *2*, it changes to a medial direction and at the inferior aspect of the cerebellum crosses
the midline (*4*), after which it curves back to the *right* side (*5*). This small shift to the opposite
side is probably due to the presence of the tumor. **B** Left vertebral angiogram, arterial phase,
lateral projection. *Arrowheads* indicate the course of the right AICA on the inferior aspect
of the cerebellum. **C** CT scan after contrast injection. A contrast-enhanced, circumscribed
mass is present in the right cerebellopontine angle

the VA were present in five other cases [C (Fig. 64), D, F (Fig. 65), G (Fig. 66),
and I (Fig. 67)]. The only case in which abnormalities in the distribution area
of the PICA and the AICA were present is B (Fig. 68). This patient presented
the clinical picture of lateral caudal pontine infarction.

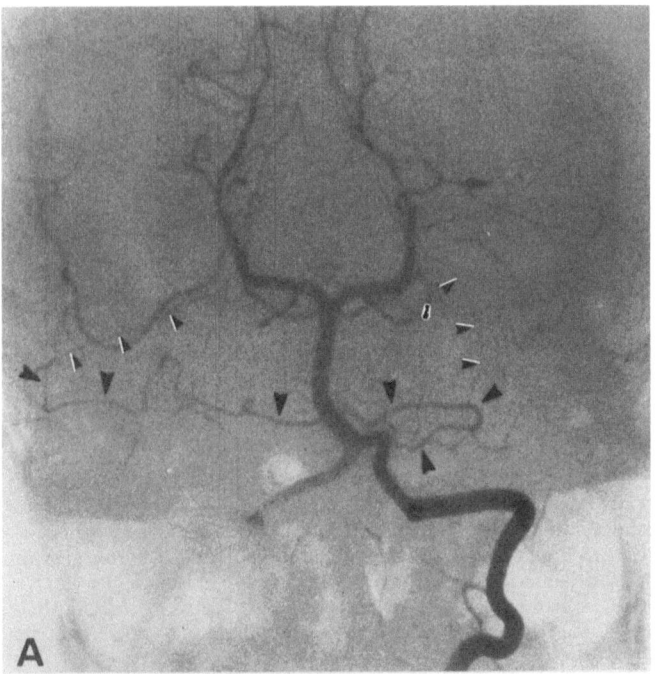

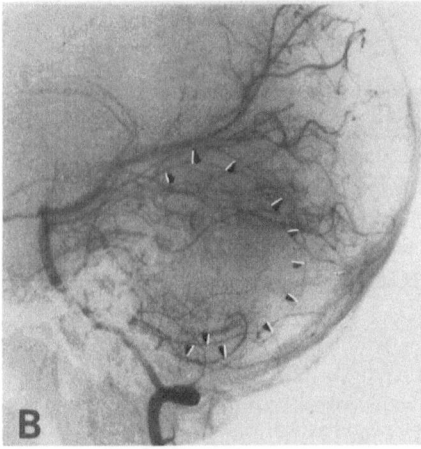

Fig. 59 A, B. Medulloblastoma in the cerebellar vermis, associated with hydrocephalus; male, aged 6 years. Left vertebral angiogram. **A** Arterial phase, half-axial projection. On the *right* side a type II AICA is present, and on the *left* side a type IV A AICA exists (*large arrow-heads*). No displacement of these vessels can be seen. Some smaller arterial branches show a stretched course (*small arrowheads*). **B** Arterial phase, lateral projection. The presence of the tumor is outlined by stretched vessels (*small arrowheads*), which are branches of the SCAs, AICAs, and right PICA

3.2.3 Discussion

In the 100 normal angiograms which were analyzed, classification of the AICA was made in the same way as in the anatomic specimens. In the majority of the angiograms, the same four main types of the artery could be distinguished.

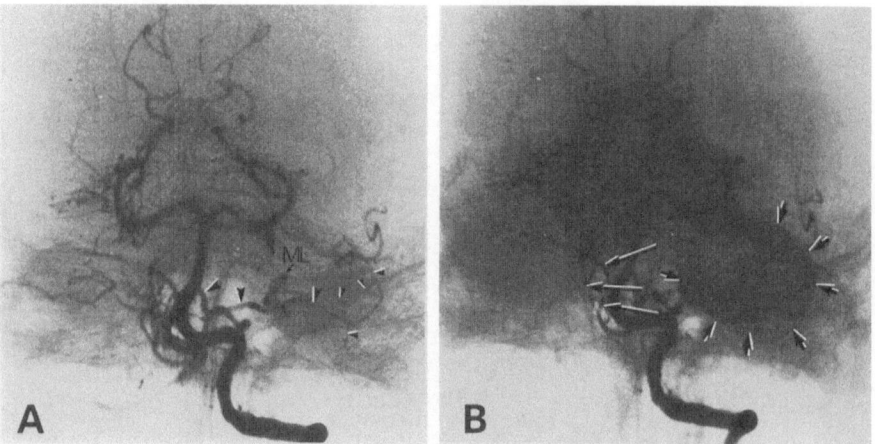

Fig. 60A, B. Tumor in the left cerebellar hemisphere, probably metastasis from a bronchial carcinoma (diagnosis not verified operatively); female, aged 71 years. Left vertebral angiogram, half-axial projection. **A** Arterial phase. A type II AICA is present on the *left* side (*large arrowheads*). The artery, including its meatal loop (*ML*), is not apparently displaced. Hemispherical branches are already visible in this phase (*small arrowheads*), but tumor staining is not yet seen. **B** Late arterial phase. The hemispherical branches of the AICA seem to contribute to the staining of a large hemispherical mass (*heavy arrows*). In this phase, the shift of the PICA (*long arrows*) to the right side is better visible than in **A**

In 6 of 14 cases which could not be classified, this was due to overprojection of AICA and PICA branches, which is a disadvantage of the half-axial or Towne projection. This disadvantage has already been pointed out by Takahashi et al. (1968, 1974) and Krayenbühl et al. (1979). As illustrated in Fig. 30, the early filling phase makes it possible to recognize the course of the AICA. When two or more AICAs of about equal size were present, no classification was made. However, in all these cases, the AICAs could be classified as one of the four types.

In the 20 anatomic specimens, the AICA was always present. This was not the case in the angiographic material: In two patients, no AICA was seen bilaterally, and in 16 patients the AICA was absent on one side. However, when an AICA is small, as in specimen 16 (Fig. 17), it can only be visualized angiographically when magnification techniques are used. It is therefore possible that in some angiograms in which no AICA could be seen, a small AICA was, in fact, present. Consequently, such an artery must be classified as a type I AICA. In this respect, it is important to note that Gabrielsen and Amundsen (1969) found that visualization of pontine arteries in the AP projection was possible in only 41% of cases. A small AICA will be of about the same caliber as such a pontine artery and, therefore, will be equally difficult to visualize. Our type II AICA is the AICA which is described in most anatomy textbooks. In our material, this type II AICA was seen in 48% of cases, which is in accordance with the observation of Takahashi et al. (1968, 1974a, b) that the AICA follows the "usual course" in about half of the cases.

In our material, a meatal loop was distinguished in the majority of cases with type II or type III AICAs, while no meatal loop was seen when the AICA

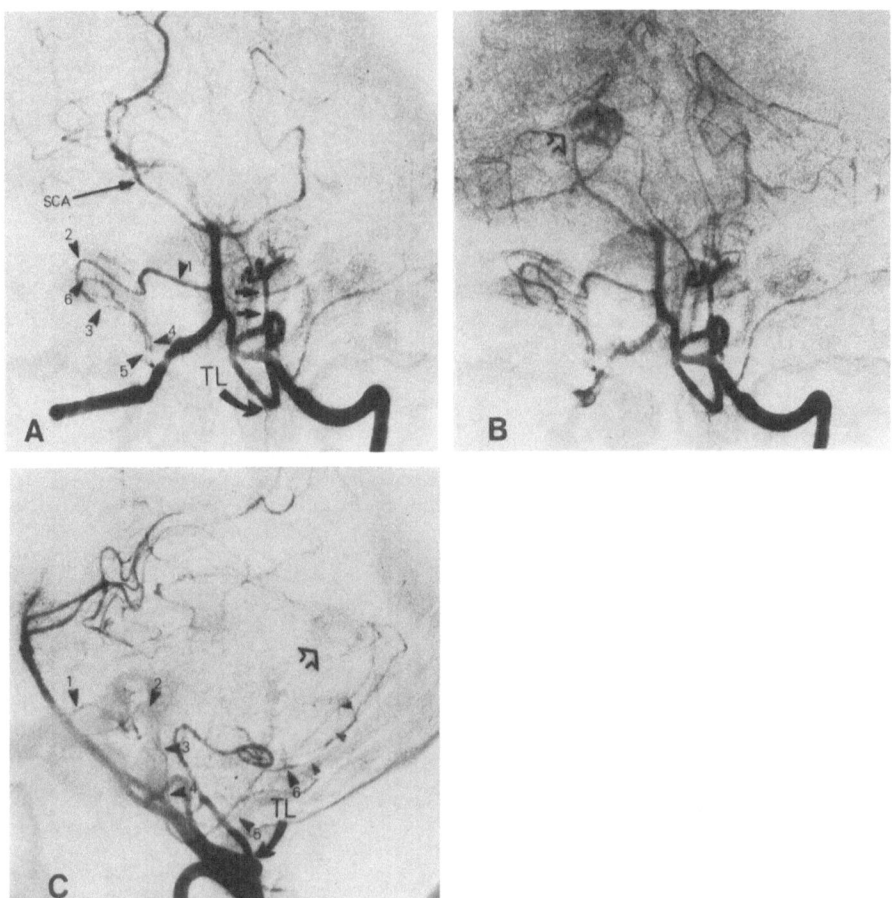

Fig. 61 A–C. Hemangioma with cyst formation in the right cerebellar hemisphere, associated with hydrocephalus; male, aged 41 years. Left vertebral angiogram. **A** Arterial phase, half-axial projection. The right SCA is displaced superiorly and medially (*long arrow*). The vermian branch of the left PICA is laterally displaced (*heavy arrows*). The tonsillar loop (*TL*) of the PICA is located below the foramen magnum. A type IV A AICA is present on the *right* side (*arrowheads*); the higher numerals are the more distal. **B** Late arterial phase, half-axial projection. A small hypervascular tumor is present in the right cerebellar hemisphere (*open arrow*). **C** Arterial phase, lateral projection. The descending course of the tonsillar loop (*TL*) of the left PICA into the foramen magnum in this projection is clearer than in the half-axial projection. This sign is not present on the *right* side, where the PICA is absent. The *numerals*, indicating the AICA (*large arrowheads*), correspond to those in the AP projection. A small hemispherical branch of the right AICA (*small arrowheads*) shows a stretched course

was absent or was of the type I or type IV configuration. In less than two-thirds of instances, a type II or type III AICA was present. Gerald et al. (1973) found a meatal loop in 60 of 74 AICAs. Therefore, the frequency of occurrence of a meatal loop in our material is lower than in the material of Gerald et al. The meatal loop was so named by Naidich et al. (1976a, b) in cases when only a single loop occurred. Where there was a double loop, that part of the artery was called the M-segment, named in reference to the shape of the loop

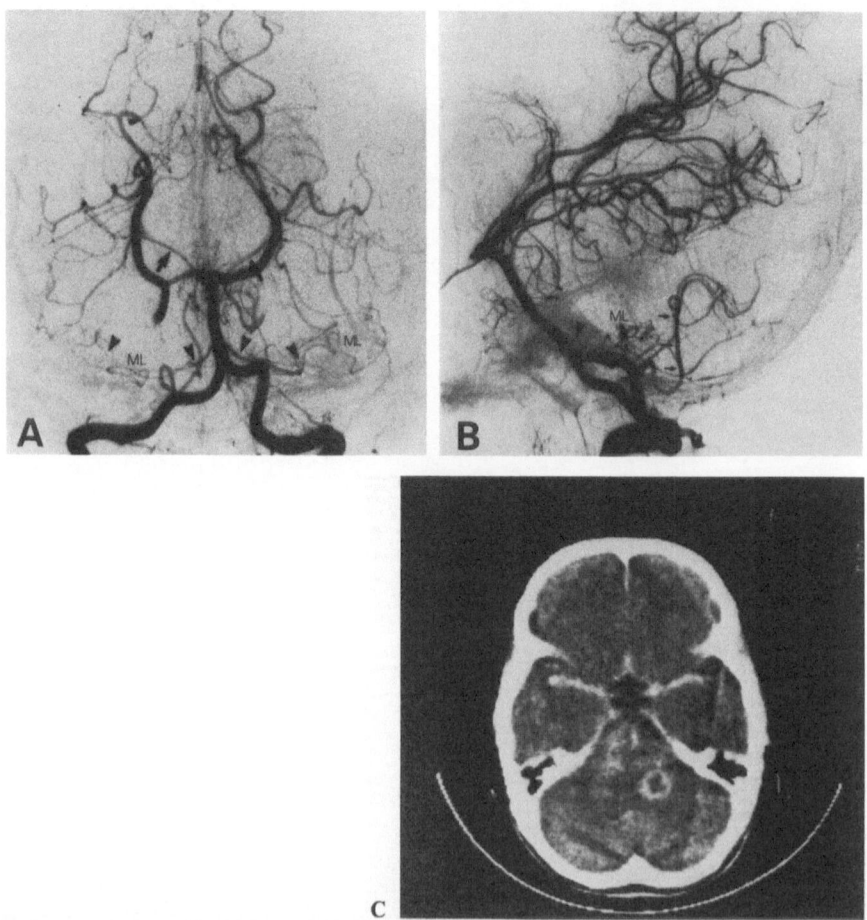

Fig.62A–C. Eccentric left-sided brain stem tumor, diagnosis not operatively verified; female, aged 4 years. **A** Left vertebral angiogram, arterial phase, half-axial projection. Both AICAs are of the type II configuration and are inferiorly displaced (*arrowheads*). The SCAs (*heavy arrows*) are superiorly and medially displaced. **B** Vertebral angiogram, arterial phase, lateral projection. The vermian branch of the right PICA is posteriorly displaced (*arrows*). The inferiorly displaced proximal parts of both AICAs are superimposed (*arrowhead*). **C** CT scan after contrast injection. A contrast-enhanced, left-sided, intra-axial lesion is present

in the lateral projection. In our material, this M-shaped configuration was present in 18 cases. There are, however, many cases in which identification of the course of this part of the artery in the lateral projection was not possible because of overprojection of both AICAs. This is illustrated in Fig. 35.

Figure 69 shows that the radiological appearance of the meatal loop in the AP projection is highly variable. The variability in the appearance of the meatal loop makes it difficult to recognize distortion of the loop by the presence of a tumor. In the report of Naidich et al. (1976a, b), minor variations in the ascending segment of the meatal loop in the presence of cerebellopontine angle tumors are mentioned, but it is also pointed out that such variations may occur when no tumor is present, especially when the artery is excessively tortuous.

Table 4. Angiographic findings in nine patients with transient or persistent ischemic lesions in the vertebrobasilar territory

Case	M/F	Age (years)	Clinical diagnosis	Angiographic findings	AICA type on side of lesion
A	F	57	Left medullary TIA	No abnormalities	III
B	M	37	Right caudal pontine infarction	PICA and AICA branches occluded	?
C	F	49	Right medullary infarction	Stenosis of right and left VA	I
D	M	33	Right lateral medullary infarction (Wallenberg's syndrome)	Occlusion of right VA	I
E	M	55	Left medullary infarction	Elongation of arteries	II
F	M	40	Left lateral medullary infarction (Wallenberg's syndrome)	Occlusion of left VA	II
G	M	46	Right pontine TIAs	Stenosis of right VA, occlusion of right PICA	II
H	F	45	Right pontine infarction	No abnormalities	IV A
I	F	59	Left medullary TIA	Stenosis of the right VA and nonunion of the VAs	II

TIA, transient ischemic attack

As already pointed out in Chap. 2, much confusion exists as regards the branching of the AICA. This is also the case in the radiological literature. In the material of Gerald et al. (1973), branching of the AICA was seen in 72 instances, while in 18 cases branching was absent. However, the site of bifurcation of the AICA was, according to Gerald et al., near the internal auditory meatus in the majority of cases. This is contrary to the findings of Naidich et al. that the AICA bifurcates into its rostrolateral and caudomedial branches near the crossing with the abducens nerve.

The case shown in Fig. 37 is an example of an AICA which bifurcates near the internal auditory meatus. In this instance, there is no medially directed branch, and therefore the AICA was classified by us as type II. An artery classified as a type III AICA bifurcates more proximally: The caudomedial branch describes a laterally convex loop and turns back toward the brain stem, thus taking over a part of the area of supply of the PICA. This is illustrated in Fig. 41.

Fig. 63. Case H, female, aged 45 years. This patient had suddenly experienced severe vertigo, nausea, and diplopia. There was cerebellar ataxia in the use of the right arm and leg. A well-sustained horizontal nystagmus appeared on both right and left lateral gaze, and rotatory nystagmus was present on upward gaze. There was some hearing loss in the right ear. Diagnosis:

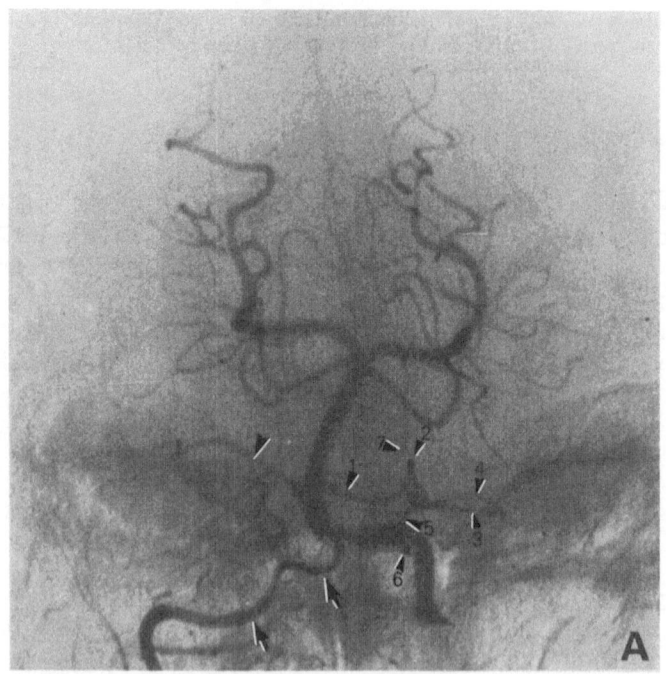

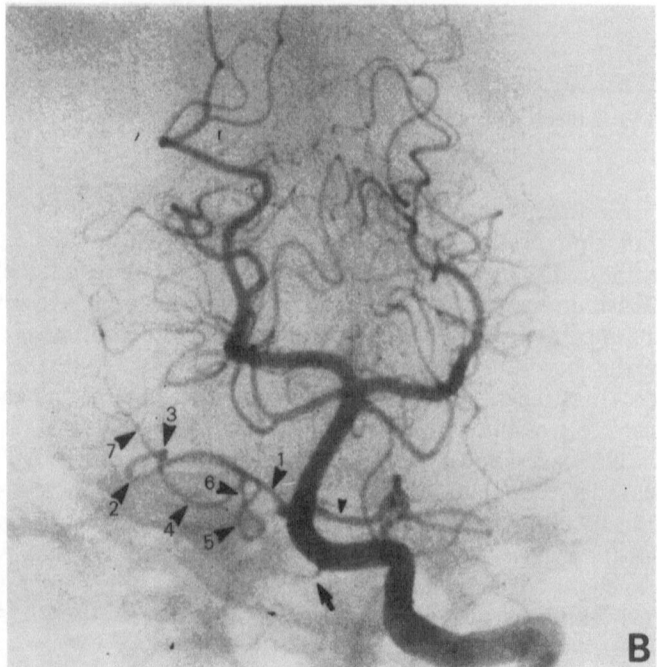

pontine infarction on the right side. **A** Right vertebral angiogram, arterial phase, half-axial projection. **B** Left vertebral angiogram, arterial phase, half-axial projection. On both sides, a type IV A AICA is present. Both AICAs are indicated by *arrowheads*, numbered *1–7*, the higher numbers being the more distal. The proximal part of the right AICA (between *1* and *2*) shows a superior convex curve, possibly a sign of swelling in the right half of the pons. The right VA is of small caliber

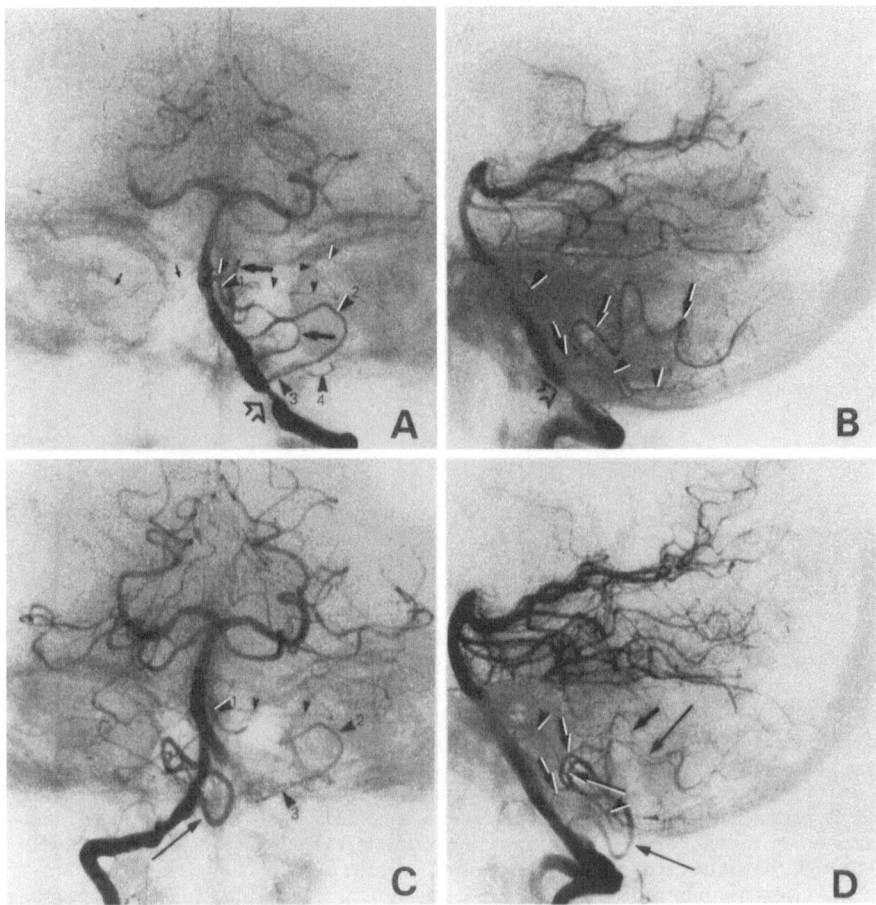

Fig. 64A–D. Case C, female, aged 49 years. The patient had experienced brief episodes of headache and dizziness during the few days prior to admission. Slurring of speech and numbness in the left arm and left leg developed suddenly. Examination showed the right corneal reflex to be hypoactive. Dysarthria and dysphagia were present. There was a moderate paresis of the left arm and the left leg. Diagnosis: right medullary infarction. **A** Left vertebral angiogram, arterial phase, half-axial projection. There is marked stenosis of the left VA, just proximal to the origin of the PICA (*heavy arrows*), which barely contributes to the blood supply of the cerebellar hemisphere. A type IV AICA is present on the *left* side (*large arrowheads*, numbered *1–4*, the higher numbers being the more distal). There is a superior accessory AICA (*small arrowheads*). The right AICA is of the type I configuration (*small arrows*). **B** Left vertebral angiogram, arterial phase, lateral projection. The origin of the left PICA (*heavy arrows*) directly distal to the stenosis of the left VA (*open arrow*) is better visible in this projection than in the AP view. The left AICA (*arrowheads*) courses on the inferior aspect of the cerebellar hemisphere. **C** Right vertebral angiogram, arterial phase, half-axial projection. The right VA shows atherosclerotic changes. The right PICA (*long arrow*) is of substantial caliber. Both the left AICA and the left accessory AICA are visible. **D** Right vertebral angiogram, arterial phase, lateral projection. Both VAs and both PICAs are filled. The right PICA extends below the foramen magnum. Symbols used are the same as in **A, B,** and **C**

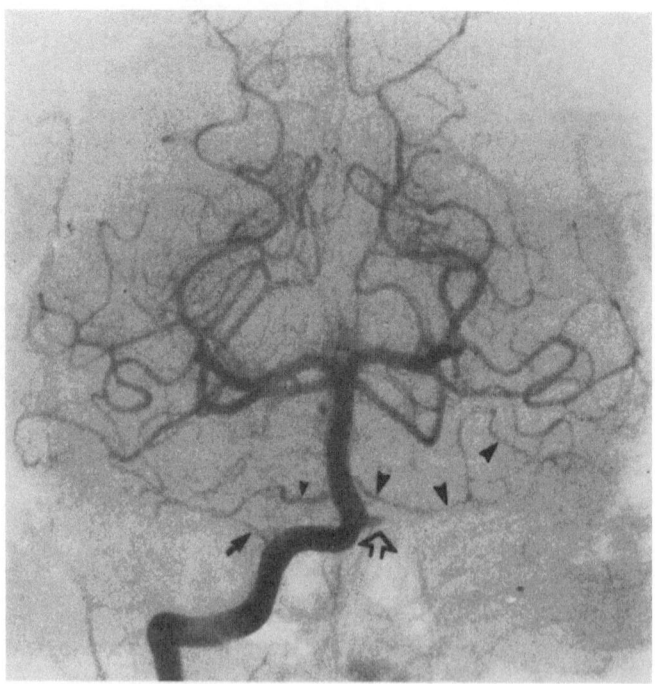

Fig. 65. Case F, male, aged 40 years. The patient had experienced episodes of occipital headache for 10 days prior to admission. During the 2 days before admission, he had suffered from persistent hiccoughs and dizziness. Examination showed the left corneal reflex to be diminished, and a left horizontal nystagmus was present on lateral gaze. A left-sided Horner's syndrome existed. A partial left peripheral facial palsy and a left-sided paresis of the pharyngeal and tongue muscles existed. Cerebellar ataxia was present in the use of the left arm and left leg. Pain and temperature sensation were diminished over the right half of the body. Diagnosis: left lateral medullary infarction. Right vertebral angiogram, arterial phase, half-axial projection. The left VA is only visible in its most distal trajectory. The *open arrow* indicates the site where it is completely occluded. No left PICA is visible. The right PICA is small (*heavy arrow*). Both AICAs are of the type II configuration and are indicated by *arrowheads*

One of the variants observed by Gerald et al. was the marginal branch of the SCA supplying the usual distribution area of the lateral branch of the AICA. In our material, this variant occurred either in cases when the ipsilateral AICA was small or absent, or when the ipsilateral AICA was of the type IV configuration.

The inverse relationship in size between AICA and PICA is emphasized by many authors. This is also reflected in our own findings. In all instances in which the AICA was absent or was of the type II configuration, the PICA was well developed. The PICA was always present in cases with a type II AICA, and in most cases with a type II AICA, the PICA was larger in caliber.

Type III A and IV A AICAs occurred more frequently than type III and IV AICAs, respectively, and this finding implies that in most cases in which the AICA had a type III or IV configuration, the PICA was absent. This finding has practical consequences for cerebellopontine angle surgery since, especially when the AICA has a large territory of supply, damage to the AICA may

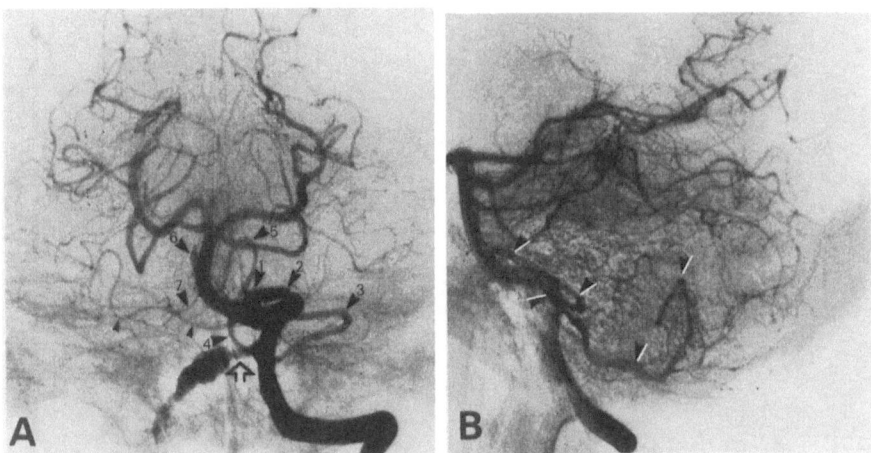

Fig. 66 A, B. Case G, male, aged 46 years. This patient had experienced sudden blurring of vision and right-sided headache 2 days prior to admission. He was admitted to hospital because of episodes of numbness in the right half of the face, associated with dysarthria and dysphagia. During these attacks, there also appeared to be a right abducens paresis, slight peripheral facial paresis on the right, and cerebellar ataxia of the right arm. Diagnosis: transient ischemic attacks of the right half of the pons. Left vertebral angiogram, arterial phase. **A** Half-axial projection. Marked tortuosity of the BA exists. There is a 50% stenosis of the right VA (*open arrow*) with aneurysmal dilatation of the part of the artery directly proximal to the stenosis. No right PICA is visible. The right AICA is of the type II configuration. On the *left* side, a type IV A AICA is present (*large arrowheads*, numbered *1–7*, the higher numbers being the more distal). The distal part of this AICA projects on the *right* side of the BA and probably supplies the caudal part of the cerebellar vermis. **B** Lateral projection. The course of the large left-sided AICA on the inferior border of the cerebellum is indicated by *arrowheads*

be followed by serious complications (Atkinson 1949). This is the case when the PICA is small or absent.

As can be seen from specimens 7 and 16 (Figs. 8, 17), the type IV AICA traverses the cerebellopontine angle cistern and then turns medially on the surface of the flocculus. As the artery reaches the fissure between the brain stem and the cerebellar hemisphere, it runs caudally and then changes course again in a lateral direction. As in these cases no PICA is present, probably a large part of the lateral medulla oblongata is supplied by branches of the AICA. If damage to a type IV AICA occurs, the larger part of the cerebellar hemisphere and possibly a part of the pons and the medulla oblongata could be infarcted.

The localizing function of vertebral angiography in the diagnosis of cerebellopontine angle tumors has been taken over by CT. Nevertheless, vertebral angiography provides information about details which are important when surgical therapy of a cerebellopontine angle tumor is considered and which are not provided by CT. Kendall and Symon (1977) point out the importance of demonstrating variations in vascular anatomy as well as the relation between the tumor and the major blood vessels. They regard vertebral angiography as particularly important in extra-axial tumors adjacent to the brain stem, such as acoustic neurinomas. Kendall and Symon also emphasize that the AICA can be displaced by a cerebellopontine angle tumor in various ways, depending

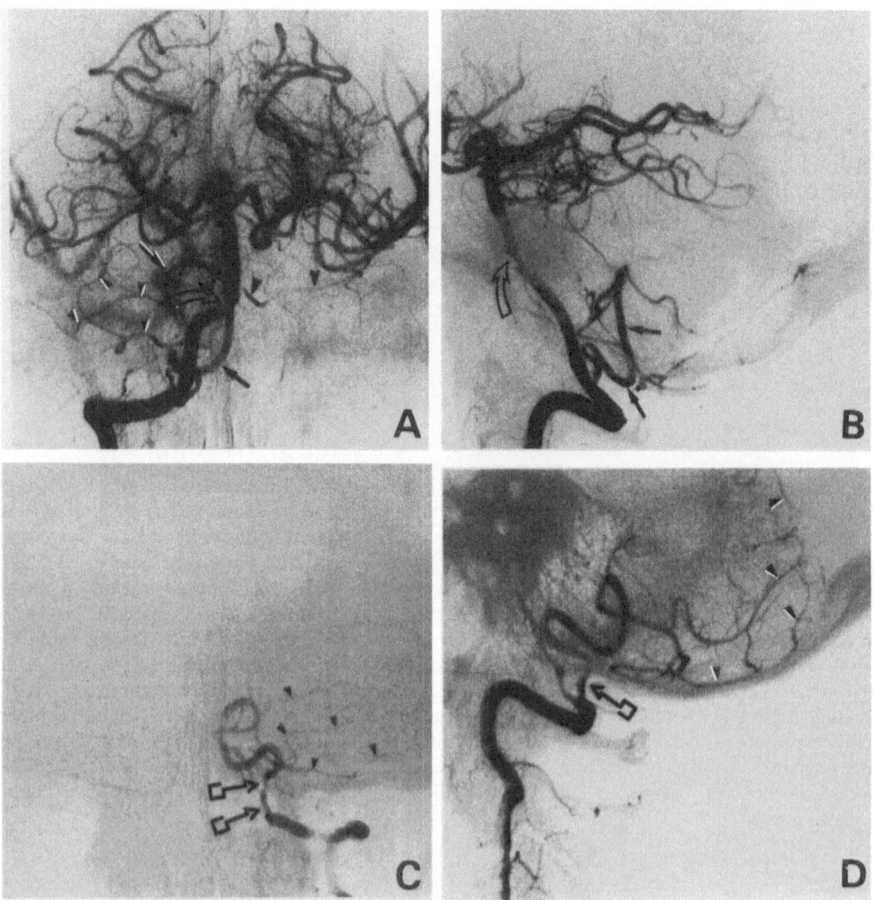

Fig. 67A–D. Case I, female, aged 59 years. This patient had experienced an episode of dysarthria, vertigo, and unsteadiness of gait, which had lasted approximately 3 h, 2 months prior to admission. She was admitted to hospital because of transient paresis of the right arm and leg. At examination, the only neurological sign was dysarthria. This improved during the first days in the hospital. Diagnosis: transient ischemic attacks in the vertebrobasilar system. **A** Right vertebral angiogram, arterial phase, half-axial projection. A stenosis of the most distal part of the right VA is present (*curved white arrow*). The large PICA (*heavy arrows*) originates a long way proximal to this stenosis. Branches of this PICA supply the right cerebellar hemisphere (*small arrowheads*). No AICA is present on the *right* side. A type II AICA is seen on the left (*large arrowheads*). **B** Right vertebral angiogram, arterial phase, lateral projection. **C** Left vertebral angiogram, arterial phase, half-axial projection. The left VA does not unite with the right VA but exclusively supplies the PICA territory (*small arrowheads*). The *arrows* indicate a stenosis of the VA at a length of about 1 cm. **D** Left vertebral angiogram, arterial phase, lateral projection

on its original relation to the vestibulocochlear nerve. Posterior, superior, and medial displacement of the AICA were the most frequent in their material. In 25% of their material, the AICA was displaced inferiorly, often combined with posterior or medial displacement. In our ten cases of cerebellopontine angle tumors with a type II AICA, similar figures were found: Superior displacement was seen in seven instances and inferior displacement in three instances.

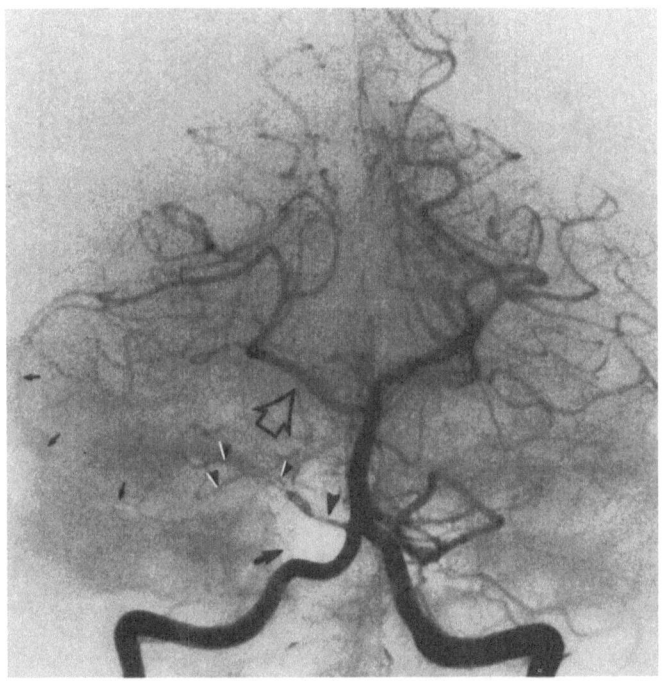

Fig. 68. Case B, male, aged 37 years. This patient experienced sudden profound dizziness and nausea, followed by deafness of the right ear and a numb sensation in the right half of the face. Examination showed the right pupil to be smaller than the left. A right-sided abducens paresis was present and a hemihypoesthesia for pain and temperature of the right side of the face was found. Furthermore, there was a right-sided facial paresis of the peripheral type and total deafness on the right side. Diagnosis: infarction of the right half of the caudal part of the pons. Left vertebral angiogram. Arterial phase. Half axial projection. There is no filling of the right posterior cerebral artery. The right SCA (*white arrow*) is extremely well-developed and its terminal branches (*small arrows*) extend to the inferior border of the right cerebellar hemisphere. At the site of junction of both VAs, an artery originates that is probably the AICA (*large arrowhead*), but shortly after its origin, only a few thin branches are present (*small arrowheads*). There is only an extremely small PICA on the right side

The combination of a superior and a medial shift was also observed (Figs. 54, 55). In the half-axial projection we routinely use, it is often difficult to differentiate between superior and posterior displacements of the AICA. According to Pinto et al. (1977), posterior displacement of the AICA by a cerebellopontine angle tumor is the most frequent. These authors point out the advantage of the base view in visualizing this type of displacement, but, in our experience, it is often impossible to position the patient such that this projection can be achieved.

The types of displacement described are especially relevant for the type II AICA. When the artery is of the type I configuration and does not reach further laterally than the internal auditory meatus, no significant displacement can be demonstrated (Fig. 50). On the other hand, possible displacement of a type IV AICA is difficult to recognize. In our opinion, however, it is especially important that the type IV configuration is recognized as such, because of the vital role the type IV AICA may play in the blood supply of the brain stem.

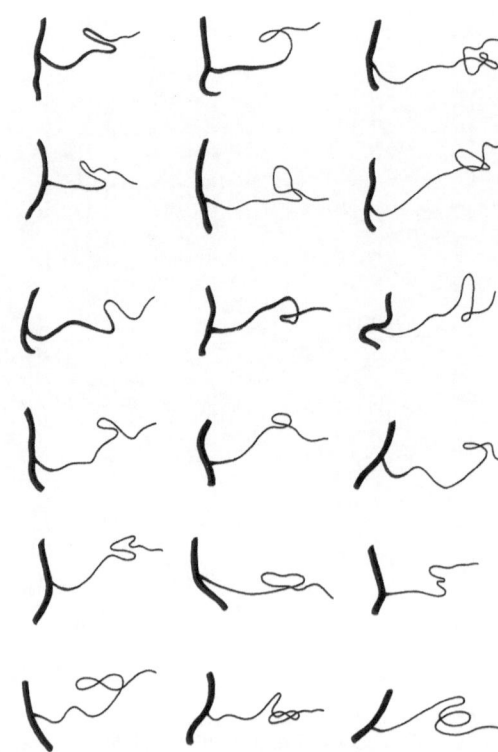

Fig. 69. Eighteen examples of a type II AICA illustrating the variability of the meatal loop. The distal branches of the AICAs are not indicated

In our material, staining of a cerebellopontine angle tumor was observed in 4 of 20 cases. This is in accordance with the findings of Ziedses des Plantes (1968), Hanafee and Wilson (1972), Takahashi et al. (1971), and Numaguchi et al. (1980). Only Kendall and Symon (1977) found a higher percentage, namely 62%.

We did not investigate the external carotid circulation in the 20 cases of cerebellopontine angle tumors. In such cases, selective external carotid injection should be considered, since there is often a contribution from branches of the external carotid artery to the vascularization of the tumor (Levine et al. 1973; Moscow and Newton 1975; Théron and Lasjaunias 1976; Kendall and Symon 1977; Perneczky 1980).

As mentioned by Kendall and Symon (1977), the increasing availability of CT has led to a decline in the use of angiography in the elucidation of posterior fossa lesions. They point out that this is not justified in the case of extra-axial tumors, such as acoustic neurinomas, but that the decline may easily be justified in the case of lesions confined to the cerebellum or brain stem, where intimate vascular anatomy is not usually of importance. In a paper on pediatric posterior fossa tumors, Rappaport and Epstein (1978) even stated that vertebral angiography is unnecessary in the preoperative evaluation of these tumors. As far as the AICA is concerned, no severe displacements were seen in the 15 cases of cerebellar tumors and in the six cases of brain stem tumors. This is in accordance with the findings of other authors (George 1974; Takahashi et al. 1974; Galanski et al. 1978), who regarded changes in the course of the AICA in the presence of cerebellar or brain stem tumors as unspecific. Only in the report of Naidich et al. (1976a, b) is a detailed description given of displacements of various segments of both main branches of the AICA in the presence of cerebellar tumors.

In the case of a medulloblastoma in the cerebellar vermis, shown in Fig. 59, the configuration of the right and left AICA can be easily recognized. Only the distal branches, supplying the vermis and the cerebellar hemisphere, show a stretched course, and in this case vertebral angiography has not provided preoperative information of vital importance. Nevertheless, vertebral angiography can reveal staining of cerebellar tumors and details of the anatomy of arteries which contribute to the blood supply of these tumors (Fig. 61).

In the case shown in Fig. 60, the hemispherical branches of the type II AICA supply the hypervascular metastasis in the left cerebellar hemisphere.

Vertebrobasilar ischemic disorders occur frequently. However, establishment of a vascular lesion in the vertebrobasilar territory often has no therapeutical consequences, and vertebral angiography carries some risks in elderly patients with vertebrobasilar vascular disease. For this reason, the number of angiograms of patients with ischemic lesions examined by us is relatively low.

In younger patients, vertebral angiography may confirm a suspected vascular lesion in the vertebrobasilar territory. A consequence of such a finding could be the prescription of antiplatelet-aggregating drugs. Operative therapy of a vascular lesion in the vertebrobasilar territory is not yet a routine procedure. However, several reports on extra-intracranial bypass operations in patients suffering from occlusive disease in the vertebrobasilar system have been published (Khodadad et al. 1977; Sundt and Piepgras 1978; Ausman et al. 1979, 1981a, b). Obviously, thorough preoperative angiographic evaluation is needed when such operations are considered.

The only case in our series of nine patients in which a suspected AICA lesion was present, was the case presented in Fig. 64. On the right side, a small AICA originated from the junction of the VAs. The ipsilateral PICA was very small, and this is unusual in the presence of an AICA of the type I configuration. Possibly, the visualized artery was an inferior accessory AICA, and the AICA itself was occluded. The clinical picture this patient presented resembles the clinical picture in the case described by Adams (1943). Adams found ipsilateral cerebellar signs, which were not present in our patient. Furthermore, Adams did not report an abducens palsy, which existed in the case described by us and which can probably be ascribed to ischemia of the abducens nerve.

In the two cases of lateral medullary infarction, vertebral angiography revealed occlusion of the ipsilateral VA. This is in accordance with the findings of Fisher et al. (1961).

Since arteries that supply the lateral medullary region originate not only from the PICA, AICA, and BA, but also from the VA, and collateral flow from one lateral medullary artery to the other is absent (Fisher et al. 1961), the lateral medullary syndrome can be caused by occlusion of the VA only. The type of AICA present on the side of the lesion is obviously not of importance for the prevention of an ischemic lesion.

82

4 SUMMARY

The aim of the present study, as set out in Chap. 1, is to provide a description of the anatomic variations of the AICA and of the radiological appearance of these variations in both normal and abnormal angiograms. Chapter 2 deals with the anatomy of the AICA. The literature on various anatomic details of the artery is reviewed, with emphasis placed upon the different criteria which have been used in naming the arteries of the posterior fossa or the various parts of these arteries.

Some authors state that the AICA and the PICA may originate from a common trunk. Others use the label AICA only for arteries which originate from the BA, irrespective of their course or area of supply. When the AICA is defined in this way, it may then possess a well-developed caudomedial branch that takes over the whole, or part of, the area of supply of the PICA. In most reports, absence of the AICA is considered rare. In contrast, multiplication of the artery is reported to occur in up to 30% of cases. In most instances, the AICA has an intimate anatomic relation to the facial and vestibulocochlear nerves. Various types of artery-nerve relations can be distinguished, depending on the configuration of the loop that the AICA describes in the vicinity of the internal auditory meatus. Several small arteries originate from this section of the AICA, one of which is the internal auditory artery. According to some authors, the internal auditory artery may also originate directly from the BA. Others, however, state that it always arises from the AICA. Obviously, these different opinions regarding the origin of the internal auditory artery depend on how the artery is defined – as an artery that exclusively supplies the inner ear, or as an artery that sends branches to the brain stem and cerebellum as well as to the inner ear.

Different criteria are used by various authors to define the main branches of the AICA, i.e. the rostrolateral or lateral branch and the caudomedial or medial branch, especially as regards the site of origin of these branches.

In an injection study of 20 normal anatomic specimens, we were able to distinguish four main types of the AICA:

1. The *type I AICA* is short and does not reach further laterally than the internal auditory meatus.
2. The *type II AICA* usually forms a so-called meatal loop, which is in close relation with the facial and vestibulocochlear nerves and with the internal auditory meatus, and then continues its course laterally in the great horizontal fissure toward the lateral border of the cerebellar hemisphere.
3. The *type III AICA* follows the same course as the type II AICA but, in addition, a caudomedial branch is present. This branch originates from the

proximal part of the AICA and takes over the whole, or part of, the area of supply of the PICA.

4. In the *type IV AICA*, no laterally coursing segment is present, and there is only a caudomedial branch.

Additionally, a type *IIIA* and a type *IVA* AICA are distinguished in cases when no ipsilateral PICA is present.

Chapter 3 is a study of the radiological appearance of the AICA. The first vertebral angiography was performed in 1933. During the 1950s and early 1960s, techniques improved, and the role of vertebral angiography in the diagnosis of posterior fossa tumors became more important. Demonstration of abnormal tumor vasculature and displacement of major arteries were both regarded as localizing signs. Adequate visualization of the AICA is only possible with sub-traction, and the first reports on the radiological significance of the AICA appeared during the late 1960s, when subtraction techniques were more con-sistently used. The importance of locating posterior fossa tumors by means of vertebral angiography has decreased drastically with the widespread use of CT scanners. Vertebral angiography is still, however, an important preoperative examination in patients with a posterior fossa tumor. Especially in cases of cerebellopontine angle tumors, severe complications of operative therapy may result from damage to the AICA. These complications are more likely to occur when the AICA has a large area of supply. The use of special projections may eventually clarify the relation between the AICA and the tumor.

One hundred normal vertebral angiograms were analyzed and the AICAs classified according to the four types described in Chap. 2. In 18 cases, no AICA was observed unilaterally or bilaterally. In some cases, this might be due to the fact that the AICA was too small to be seen without magnification. The type I AICA was present in 9.5% of cases. The type II AICA occurred the most frequently and was seen in almost half of the cases investigated. In the majority of these cases, a meatal loop was present. The radiological appearance of this loop is highly variable. Types IIIA and IVA AICAs were more frequent than types III and IV, respectively. This implies that the PICA is often absent when the caudomedial branch is strongly developed.

In addition to the 100 normal vertebral angiograms, 41 angiograms from patients with posterior fossa tumors were analyzed. In tumors of the cerebellar hemisphere, cerebellar vermis, or brain stem, the AICA was not usually signifi-cantly displaced. In the 20 patients suffering from a cerebellopontine angle tu-mor, the AICA was not clearly displaced when the artery was of the type I configuration. The type II AICA can show either a superior or an inferior displacement, sometimes in combination with a posterior or a medial shift. In the material reviewed, no cases of cerebellopontine angle tumors with a type III or IIIA AICA were present.

It is often not possible to recognize a displacement of a type IV AICA, but since this type of the artery is frequently seen in combination with a small or absent PICA, it is important to recognize this configuration, because in these cases the AICA plays an important role in the vascularization of the brain stem and cerebellum. The same is true for the type III AICA.

Nine angiograms of patients with transient or persistent ischemia in the vertebrobasilar territory were analyzed. One of these patients showed the clinical picture of lateral pontine infarction. In this patient, both the AICA and the

PICA on the ipsilateral side were small, and probably one of the major AICA branches was occluded. In the other eight patients, various lesions of the vertebral arteries could be demonstrated, but no lesions of the AICAs were present.

Acknowledgements. We are indebted to Mrs. M. Stiaszny and J.H.A. van der Drift for reviewing the manuscript.

We are grateful to Mrs. A. Bredius for her valuable assistance and to Mrs. E. Zijlstra for typing the manuscript.

We express our appreciation to L. Cremers, C. van Ederen, C. van der Hout, D. de Jong, J.F. Olijhoek, F. ter Riet, and S.H. Speelman for technical assistance.

5 REFERENCES

Adachi B (1928) Das Arteriensystem der Japaner, vol 1. Verlag der Kaiserl Jap Univ Kyoto. p 119

Adams RD (1943) Occlusion of the anterior inferior cerebellar artery. Arch Neurol Psychiat 49:765–770

Adams RD, Victor M (1977) Principles of Neurology, McGraw-Hill, New York, pp 512–518

Alezaïs MM, d'Astros L (1892) La circulation artérielle du pédoncule cérébral. J Anat (Paris) 28:519–528

Allcock JM (1962) Vertebral angiography: its accuracy in the diagnosis of intracranial space-occupying lesions. J Can Assoc Radiol 13:65–69

Amplatz K, Harner R (1962) A new subclavian artery catheterization technic. Radiology 78:963–966

Atkinson WJ (1949) The anterior inferior cerebellar artery. J Neurol Neurosurg Psychiatry 12:137–151

Ausman JI, Lee MC, Chater N, Latchaw RE (1979) Superficial temporal artery to superior cerebellar artery anastomosis for distal basilar artery stenosis. Surg Neurol 12:277–282

Ausman JI, Diaz FG, Reyes RA de los, Pak H, Patel S, Boulos R (1981a) Superficial temporal to proximal superior cerebellar artery anastomosis for basilar artery stenosis, Neurosurgery 9:56–59

Ausman JI, Diaz FG, Reyes RA de los, Patel S (1981b) Anastomosis of occipital artery to anterior inferior cerebellar artery for vertebrobasilar junction stenosis. Surg Neurol 16:99–102

Bender MB (1973) Cerebellopontine angle tumors or acoustic neuromas: long-range management. Arch Otolaryngol 97:160–165

Berczeller A, Kugler H (1937) Freilegung der Arteria vertebralis am Sulcus atlantis. Beitrag zur Arteriographie des Stromgebietes der Arteria vertebralis basilaris. Arch Klin Chir 190:810–815

Berg D van den, Drift JHA van der (1963) Vertebro-basilaire angiografie via catheterisatie van de arteria brachialis. Ned Tijdschr Geneeskd 39:1743–1748

Berry RJA, Andersen JH (1910) A case of non-union of the vertebrals, with consequent abnormal origin of the basilar. Anat Anz 35:54–65

Blackburn JW (1907) Anomalies of the encephalic arteries among the insane. J Comp Neurol Psychol 17:493–517

Böhne C (1927) Über die arterielle Blutversorgung des Pons. Z Anat Entwickl Gesch 84:777–786

Bonte G, Riff G, Spy E (1958) Angiographie vertébrale par cathéterisme retrograde fémoral. Acta Radiol 50:67–76

Braun W, Lotze M, Tänzer A (1966) Vertebralis Angiographie mit Hilfe des Femoraliskatheters. Fortschr Röntgenstr 104:839–847

Brenner H (1961) Zur diagnostischen Bedeutung der Vertebralisangiographie. Wien Klin Wochenschr 73:419–421

Brouckaert L, Sieben G, Reuck J de, Vandereecken H (1981) The syndrome of the anterior inferior cerebellar artery. Acta Neurol Belg 81:65–73

Castaigne P, Pertuiset B, Cambier J, Brunet P (1967) Aneurisme de l'artère auditive interne révélé par une paralysie faciale récidivante. La Presse Médicale 75:2493–2496

Cavatorti P (1908) Il tipo normale e le variazione delle arterie della base dell'encephalo. Monit Zool Ital 10:248–259

Charachon R, Latarjet M (1962) Les injections de matières plastiques appliquées à l'étude de l'artère auditive interne. CR Ass Anat 48:436–441

Collins WF Jr, Slade HW, Lockhart WG (1957) Brachial vertebral angiography in adults. J Neurosurg 14:466–468

Columella F, Papo I (1955) Unsere Erfahrungen mit der Vertebralisangiographie (Bericht über 220 Fälle). Zentralbl Neurochir 15:294–301

Critchley M, Schuster P (1933) Beiträge zur Anatomie und Pathologie des Arteria cerebelli superior. Z Ges Neurol Psychiat 144:681–741

Cronquist S (1961) Vertebral catheterization via the femoral artery. Acta Radiol, 55:113–118

Cushing H (1910) Strangulation of the nervi abducentes by lateral branches of the basilar artery in cases of brain tumor. Brain 33:204–235

Dilenge D, David M (1967) L'angiographie vertébrale. Neurochirurgie 13:121–156

Fazzari I (1929) Die Arterien des Kleinhirns. Anat Anz 67:497–501

Ferrari Lelli G (1939) Comportamento dell'arteria uditiva interna e dei suoi rami labirintici nell'uomo. Z Anat Entwickl Gesch 110:48–80

Fisch U (1969) The surgical anatomy of the so-called internal auditory artery. Nobel Symp Serie 10:121–130

Fisher CM, Karnes WE, Kubik CS (1961) Lateral medullary infarction – the pattern of vascular occlusion. J Neuropathol Exp Neurol 20:323–379

Foix C, Hillemand P, Schalit I (1925a) Irrigation du bulbe. C R Soc Biol (Paris) 92:33–35

Foix C, Hillemand P, Schalit I (1925b) Sur le syndrome latéral du bulbe et l'irrigation du bulbe supérieur. Rev Neurol (Paris) 32:160–179

Fujii K, Lenkey C, Rhoton AL Jr (1980) Microsurgical anatomy of the choroidal arteries. J Neurosurg 52:502–524

Gabrielsen TO, Amundsen P (1969) The pontine arteries in vertebral angiography. Am J Roentgenol 106:296–302

Galanski M, Friedman G, Lanckohr H (1978) Angiographische Befunde bei 86 gesicherten Tumoren der hinteren Schädelgrube. Unter besonderer Berücksichtigung der Lokalisation und des jeweils charakteristischen Gefässverlaufs. Fortschr Röntgenstr 128:16–23

George AE (1974) A systemic approach to the interpretation of posterior fossa angiography. Radiol Clin North Am 12:371–400

Gerald B, Wolpert SM, Haimovici H (1973) Angiographic anatomy of the anterior inferior cerebellar artery. Am J Roentgenol 118:617–621

Gillilan LA (1964) The correlation of the blood supply to the human brain stem with clinical brain stem lesions. J Neuropathol Exp Neurol 23:78–108

Gillilan LA (1969) The arterial and venous blood supplies to the cerebellum of primates. J Neuropathol Exp Neurol 28:295–307

Goodhart S, Davison C (1936) Syndrome of the posterior inferior and anterior inferior cerebellar arteries and their branches. Arch Neurol Psychiat 35:501–524

Goree DA, Tindall GT Odom GI (1964) Percutaneous retrograde brachial angiography in the diagnosis of acoustic neurinoma. Am J Roentgenol 92:829–835

Greitz T, Lindgren E (1971) The Head. In: HL Abrams (ed), Angiography, 2nd edn: vol 1. Little, Brown and Company, Boston, pp 155–281

Greitz T, Sjögren SE (1963) The posterior inferior cerebellar artery. Acta Radiol [Diagn] (Stockh) 1:284–297

Guerrier Y, Villacèque G (1949) Origine et comportement des artères cérébelleuse moyenne et auditive interne. C R Ass Anat 36:377–382

Hakuba A (1977) Anatomical relationships between the anterior inferior cerebellar artery and the facial nerve, and its surgical significance in cerebellopontine angle tumors. Neurol Surg 5:1143–1150

Hanafee WN (1963) Axillary artery approach to carotid, vertebral, abdominal aorta and coronary angiography. Radiology 81:559–567

Hanafee WN, Shinno JM (1966) Second-order subtraction with simultaneous bilateral carotid, internal carotid injections. Radiology 86:334–341

Hanafee WN, Wilson GH (1972) Head and neck. In: Hanafee WN (ed). Selective angiography. Williams and Wilkins, Baltimore, pp 113–140. (Golden's diagnostic radiology, section 18)

Hayman LA, Evans RA, Harrell JE (1979) Cerebellopontine angle tumor: diagnosis by oblique projection vertebral angiography. Radiology 130:383–385

Hiller F (1952) The vascular syndromes of the basilar and vertebral arteries and their branches. J Nerv Ment Dis 116:988–1016

Hitselberger WE, House WF (1966) Acoustic tumour surgery. Arch Otolaryngol 84:255–260

Hoffmann GR, Leifer C (1967) Réflexions sur l'intérêt de l'artériographie vertébrale dans le diagnostic des tumeurs des hémisphères cérébelleux. Ann Radiol (Paris) 10:833–836

Huang YP, Wolf BS (1970) Differential diagnosis of fourth ventricle tumors from brain stem tumors in angiography. Neuroradiology 1:4–19

Isfort A (1960) Gutartige Tumoren im Vertebralisangiogramm. Fortschr Röntgenstr 92:676–689

Ito J, Takeda N, Suzuki Y (1980) Anomalous origin of the anterior inferior cerebellar arteries from the internal carotid artery. Neuroradiology 19:105–109

Jakob A (1928) Das Kleinhirn. In: Möllendorff W von (ed): Handbuch der mikroskopischen Anatomie des Menschen, vol 4, part 1. Springer (Berlin)

Johnson JH Jr, Kline DG (1978) Anterior inferior artery aneurysms: case report. J Neurosurg 48:455–460

Kaplan HA (1956) Arteries of the brain; an anatomical study. Acta Radiol (Stockh) 46:364–370

Kaplan HA (1959) Vascular anatomy and syndromes of the midbrain and hindbrain. Clin Neurosurg 6:168–176

Kendall B, Symon L (1977) Investigation of patients presenting with cerebellopontine angle syndromes. Neuroradiology 13:65–84

Kerber CW, Margolis MT, Newton TH (1972) Tortuous vertebrobasilar system: cause of cranial nerve signs. Neuroradiology 4:74–77

Khilnani M, Silverstein A (1963) Displacement of the superior cerebellar artery. Arch Neurol 8:52–55

Khodadad G, Singh RS, Olinger CP (1977) Possible prevention of brain stem stroke by microvascular anastomosis in the vertebrobasilar system. Stroke 8:316–321

Kieffer SA, Binet EF, Gold LHA (1975) Angiographic diagnosis of intra- and extra-axial tumors in the cerebellopontine angle. Am J Roentgenol 124:297–309

Konaschko PI (1927) Die Arteria auditiva interna des Menschen und ihre Labyrinthäste. Z Anat Entwickl Gesch 83:241–268

Krayenbühl H, Yasargil MG (1957) Die vaskulären Erkrankungen im Gebiet der Arteria vertebralis und Arteria basialis. Thieme, Stuttgart, p 170

Krayenbühl H, Yasargil MG, Huber P (1979) Zerebrale Angiographie für Klinik und Praxis, 3rd edn. Thieme, Stuttgart, p 386

Lazorthes G (1961) Vascularisation et circulation cérébrales. Masson et Cie, Paris, p 323

Lazorthes G, Poulhès J, Espagno J (1950a) Les artères du cervelet. C R Ass Anat, 62:279–288

Lazorthes G, Poulhès J, Espagno J (1950) Les territoires vasculaires du cortex cérébelleux. C R Ass Anat 62:289–297

Lazorthes G, Gouazé A, Salamon G (1976) Vascularisation et circulation de l'encéphale. Tome premier: Anatomie descriptive et fonctionnelle. Masson et Cie, Paris, p 322

Lefèbvre J, Fauré C, Salamon G (1963) Étude radiologique des gliomes infiltrants du tronc cérébral. Acta Radiol [Diagn] (Stockh) 1:343–357

Leman P, Cohadon F, Leifer C (1967a) Descriptions des trajets normaux des artères de la fosse postérieure. Ann Radiol (Paris) 10:781–790

Leman P, Cohadon F, Leifer C (1967b) Valeur de l'artériographie vertébrale dans les tumeurs de l'angle pontocérébelleux. Ann Radiol (Paris) 10:791–802

Leman P, Cohadon F, Leifer C (1967c) L'artériographie vertébrale dans les neurinomes du nerf auditif. Neurochirurgie 13:752–761

Levine HL, Ferris EJ, Spatz EL (1973) External carotid blood supply to acoustic neuromas: report of two cases. J Neurosurg 38:516–520

Lindgren E (1950) Percutaneous angiography of the vertebral artery. Acta Radiol 33:389–404

Lindgren E (1956) Another method of vertebral angiography. Acta Radiol 46:257–261

Löfgren O (1956) Vertebral angiography in the diagnosis of hydrocephalus and differentiation between stenosis of the aquaduct and cerebellar tumor. Acta Radiol 46:187–194

Long JM, Kier EL, Schechter MM (1973) Radiology of epidermoid tumors of cerebellopontine angle. Neuradiology 6:188–192

Maillot C, Koritke JG, Laude M (1976) La vascularisation de la toile choroïdienne inférieure chez l'homme. Arch Anat Histol Embryol (Strasb) 59:33–70

Malter IJ, Roberson G (1972) Angiographic demonstration of anterior inferior cerebellar artery aneurysm by use of the base view. Radiology 103:125–126

Mani RL, Newton TH, Glickman MG (1968) The superior cerebellar artery: an anatomic-roentgenographic correlation. Radiology 91:1102–1108

Martin RG, Grant JL, Peace D, Theiss C, Rhoton AL (1980) Microsurgical relationships of the anterior inferior cerebellar artery and the facial-vestibulocochlear nerve complex. Neurosurgery 6:483–507

Mazzoni A (1969) Internal auditory canal. Arterial relations at the porus acusticus. Ann Otol Rhinol Laryngol 78:797–814

Mazzoni A, Hansen CC (1970) Surgical anatomy of the arteries of the internal auditory canal. Arch Otolaryngol 91:128–135

McMinn RMH (1953) A case of non-union of the vertebral arteries. Anat Rec 116:283–286

Mitterwallner F von (1955) Variationsstatistische Untersuchungen an den basalen Hirngefässen. Acta Anat (Basel) 24:51–88

Mones R (1961) Vertebral angiography: an analysis of 106 cases. Radiology 76:230–236

Moniz E, Alves A (1933) L'importance diagnostique de l'artériographie de la fosse postérieure. Rev Neurol (Paris) 2:91–96

Moniz E, Pinto A, Alves A (1933) Artériographie du cervelet et des autres organes de la fosse postérieure. Bull Acad Natl Med (Paris) 109:758–760

Moscow NP, Newton TH (1975) Angiographic features of hypervascular neurinomas of the head and neck. Radiology 114:635–640

Nabeya D (1923) A study in the comparative anatomy of the bloodvascular system of the internal ear in mammalia and in homo (in Japanese). Acta Sch Med Univ Kyoto 6:1–128

Nager GT (1954) Origins and relations of the internal auditory artery and the subarcuate artery. Ann Otol Rhinol Laryngol 63:51–61

Naidich TP, Kricheff II, George AE, Lin JP (1976a) The normal anterior inferior cerebellar artery. Radiology 119:355–373

Naidich TP, Kricheff II, George AE, Lin JP (1976b) The anterior inferior cerebellar artery in mass lesions. Radiology 119:375–383

Newton TH, Kramer RA, Mani JR (1966) Catheter technic in vertebral angiography. Radiology 87:691–695

Numaguchi Y, Kishikawa T, Ikeda J (1980) Angiographic diagnosis of acoustic neurinomas and meningiomas in the cerebellopontine angle – a reappraisal. Neuroradiology 19:73–80

Olivecrona H (1935) Bericht über arteriographische Darstellung der A. vertebralis. Zentralbl Chir 32:1904–1905

Olivecrona H (1967) Acoustic tumors. J Neurosurg 26:6–13

Olsson O (1953a) Vertebral angiography in the diagnosis of acoustic nerve tumors. Acta Radiol 39:265–272

Olsson O (1953b) Vertebral angiography. Acta Radiol 40:103–105

Peeters FLM (1968) Angiography of the vertebral artery in the diagnosis of tumors of the pons cerebelli. Radiol Clin (Basel) 37:89–93

Peeters FLM (1969) Het vertebralis angiogram bij intra-craniële tumoren. Thesis University of Amsterdam, p 166

Perneczky A (1980) Blood supply of acoustic neurinomas. Acta Neurochir (Wien) 52:209–218

Perneczky A, Tschabitscher M (1976) Kleinhirnarteriensegmente. Verh Anat Ges 70:385–391

Perneczky A, Perneczky G, Tschabitscher M, Samec P (1981) The relationship between the caudolateral pontine syndrome and the anterior inferior cerebellar artery. Acta Neurochir (Wien) 58:245–257

Pertuiset B, Nachanakian A, Gardeur D, Yacoubi A, Ancri D, Metzger J, Kujas M (1979) Protocole d'exploration des tumeurs cérébrales sustentorielles. Neurochirurgie 25:11–18

Philippon J, Gardeur D, Nachanakian A, Metzger J (1979) Approche diagnostique pré-opératoire des tumeurs de la fosse postérieure de l'adulte. Neurochirurgie 25:139–146

Pickering EC (1904) Seventy-six new variable stars. Circulars 79:1

Pinto RS, George AE, Kricheff II, Naidich TP, Fox A (1977) The base view in vertebral angiography. Radiology 124:157–164

Pygott F, Hutton CF (1959) Vertebral arteriography by percutaneous brachial artery catheterisation. Br J Radiol 32:114–119

Radner S (1947) Intracranial angiography via the vertebral artery. Acta Radiol 28:838–842

Radner S (1951) Vertebral angiography by catheterization. Acta Radiol, Suppl 87

Rappaport ZH, Epstein F (1978) Computerized axial tomography. The only preoperative study for pediatric posterior fossa tumors. Neurocirugia (Santiago) 36:5–11

Ross P, Boulay GH du (1976) An atlas of normal vertebral angiograms. Butterworths, London Boston, p 126

Roy P (1965) Percutaneous catherization via the axillary artery. Am J Roentgenol 94:1–18

Salamon G, Huang YP (1976) Radiologic anatomy of the brain. Springer, Berlin Heidelberg New York, p 282

Sartor K (1976) Einführung in die Neuroradiologie. Witzstrock, Baden-Baden, p 359

Scatliff JA, Mishkin MM, Hyde J (1965) Vertebral arteriography: an evaluation of methods. Radiology 85:14–21

Scotti G (1975) Anterior inferior cerebellar artery originating from the cavernous portion of the internal carotid artery. Radiology 116:93–94

Seeger JF, Gabrielsen TO (1972) Angiography of eccentric brain stem tumors. Radiology 105:343–351

Seldinger SI (1953) Catheter replacement of the needle in percutaneous arteriography. A new technique. Acta Radiol 39:368–384

Shalit MN, Reichenthal E (1978) Anomalous anterior inferior cerebellar artery simulating intracanalicular acoustic nerve tumor. Surg Neurol 10:337–339

Shimidzu K (1937) Beiträge zur Arteriographie des Gehirns; eine einfache perkutane Methode. Arch Klin Chir 188:295–316

Sjöquist O (1938) Arteriographische Darstellung der Gefässe der hinteren Schädelgrube. Chirurg 10:377–380

Smaltino F, Bernini FP, Elefante R (1971) Normal and pathological findings of the angiographic examination of the internal auditory artery. Neuroradiology 2:216–222

Stopford JSB (1916a) The arteries of the pons and medulla oblongata. J Anat Phys 50:131–164

Stopford JSB (1916b) The arteries of the pons and medulla oblongata – part II. J Anat Phys 50:255–280

Sugar O, Holden LB, Powell CB (1949) Vertebral angiography. Am J Roentgenol 61:166–182

Sunderland S (1945) The arterial relations of the internal auditory meatus. Brain 68:23–27

Sunderland S (1948) Neurovascular relations and anomalies at the base of the brain. J Neurol Neurosurg Psychiat 11:243–257

Sundt TM Jr, Piepgras DG (1978) Occipital to posterior inferior cerebellar artery bypass surgery. J Neurosurg 48:916–928

Takahashi K (1940) Die perkutane Arteriographie der Arteria vertebralis und ihrer Versorgungsgebiete. Arch Psychiat Nervenkr 111:373–379

Takahashi M (1974a) The anterior inferior cerebellar artery. In: Newton TH, Potts DG (eds), Radiology of the skull and brain, vol 2, Book 2, Chap 70. Mosby, Saint Louis

Takahashi M (1974b) Atlas of vertebral angiography. Urban and Schwarzenberg, München, p 384

Takahashi M (1974c) Evaluation of small branches of the vertebrobasilar system by magnification vertebral angiography. Nippon Acta Radiol 34:479–484

Takahashi M, Wilson G, Hanafee WN (1968) The anterior inferior cerebellar artery: its radiographic anatomy and significance in the diagnosis of extra-axial tumours of the posterior fossa. Radiology 90:281–287

Takahashi M, Okudera T, Tomanaga M, Kitamura K (1971) Angiographic diagnosis of acoustic neurinomas: Analysis of 30 lesions. Neuroradiology 2:191–200

Tatelman M, Sheehan S (1962) Total vertebral-basilar arteriography via transbrachial catheterization. Radiology 78:919–929

Taveras JM, Wood EH (1976) Diagnostic neuroradiology, 2nd edn: vol 2. Williams and Wilkins, Baltimore, p 708

Théron J, Lasjaunias P (1976) Participation of the external and internal carotid arteries in the blood supply of acoustic neurinomas. Radiology 118:83–88

Tönnis W, Schiefer W (1959) Zirkulationsstörungen des Gehirns im Serienangiogramm. Springer, Berlin Heidelberg New York

Tschabitscher M, Perneczky A (1974) Über die Beziehung von Kleinhirnarterien zum Meatus acusticus internus. Acta Anat (Basel) 88:231–244

Tschabitscher M, Perneczky A, Sinzinger H, Doskar K, Hoyer Ch (1975) Untersuchungen über die Gefäszversorgung des Kleinhirns beim Menschen. Verh Anat Ges 69:559–561

Tschernyscheff A, Grigorowsky I (1929) Über die arterielle Versorgung des Kleinhirns. Arch Psychiat Nervenkr 89:482–569

Valavanis A, Dabir K, Oguz M, Wellauer J (1982) The current state of the radiological diagnosis of acoustic neuroma. Neuroradiology 23:7–13

Valvassori GE (1969) Abnormal internal auditory canal: diagnosis of acoustic neuroma. Radiology 92:449–459

Vogelsang H (1974) Angiographische Studie zur Darstellbarkeit von Arterien im Meatus acusticus internus (A. cerebelli labyrinthi und A. auditiva int). Radiologe 14:548

Walker EA (1965) The vertebro-basilar arterial system and internal auditory angiography. Laryngoscope 75:369–407

Watt JC, McKillop AN (1935) Relation of arteries to roots of nerves in posterior cranial fossa in man. Arch Surg 30:336–345

Weibel J (1966) Angiography of the vertebrobasilar arterial system. Acta Radiol [Diagn] (Stockh) 5:570–580

Wende S, Nakayama N (1972) Die neuroradiologische Diagnostik des Kleinhirnbrückenwinkeltumors. Z Neurol 203:1–12

Wolf BS, Newman CM, Khilnani M (1962) The posterior inferior cerebellar artery on vertebral angiography. Am J Roentgenol 87:322–337

Wolpert S (1971) Angiography in posterior fossa tumors of infancy and childhood. Am J Roentgenol 112:296–305

Yasargil MG (1962) Die Vertebralisangiographie. Ihre Bedeutung für die Diagnose der Tumoren. Springer, Vienna

Ziedses des Plantes BG (1934) Planigraphie en subtractie. Röntgenographische differentiatiemethoden. Thesis, University of Utrecht

Ziedses des Plantes, BG (1961) Subtraktion. Thieme, Stuttgart, p 72

Ziedses des Plantes, BG (1968) X-ray examination in cerebellopontine angle tumours. Psychiat Neurol Neurochir 71:133–139

6 SUBJECT INDEX

L. Heimer

The Human Brain and Spinal Cord

Functional Neuroanatomy and Dissection Guide

1983. 213 figures, mostly in color. XI, 402 pages
ISBN 3-540-90741-6

The Human Brain and Spinal Cord is a concise introduction
to neuroanatomy and neuroscience written by one of the
most respected neuroanatomy educators in the US. The text
has been developed and refined from the author's teaching
experiences in both the US and Europe. For the first time, a
meticulously illustrated dissection guide is included, coordi-
nated to the functional anatomy text to save students the
expense and trouble of buying an additional neuroanatomy
atlas. The book is divided into four parts. Part 1 explains
terminology, brain and spinal cord development, and the
meninges and cerebrospinal fluid. Part 2 is the dissection
guide. Found here are outstanding, detailed illustrations
which were created especially for this book. Text and illustrations are arranged to maximize the book's usefulness in
the dissecting room. Part 3 is an illustrated account of functional anatomy of the brain and spinal cord, with schematic
drawings based on the illustrations found in Part 2. Part 4
covers the blood supply to the brain and spinal cord. An
appendix on peripheral nerves is also included. Clinical
correlations and case studies are provided throughout the
book to emphasize the clinical relevance of the information.

Techniques in Neuroanatomical Research

Editors: **C. Heym, W.-G. Forssmann**

1981. 165 figures. XIII, 395 pages
ISBN 3-540-10686-3

Techniques in Neuroanatomical Research is a detailed
presentation of selected methods the neurobiologist needs
for conducting morphological studies. The formulae and
practical introductions provided for each technique allow
even non-morphologists to acquire them easily. Each chapter concludes with an extensive bibliography to enable the
reader to broaden his knowledge while at the same time
introducing him to the works of leading scientists in the
field.

Springer-Verlag
Berlin
Heidelberg
New York
Tokyo

Advances in Anatomy, Embryology and Cell Biology

Editors: F. Beck, W. Hild, R. Ortmann, J. E. Pauly, T. H. Schiebler

Springer-Verlag
Berlin
Heidelberg
New York
Tokyo